THE SOCIAL VALUE OF ZOOS

Combining anecdotes with scientific data, this book is a journalistic inquiry into what is currently known about zoos and aquariums as sociocultural intersections of mission, public perception, and on-site meaning making. The authors draw on conservation psychology and other social science research to explore how zoos might develop and deliver more effective learning experiences to promote and nurture conservation values and collective action. While people use zoos with specific priorities and motivations in mind, these are social settings. Indeed, it is because they represent an important, vast, and trusted social enterprise that zoos have such powerful opportunities to change how diverse public audiences view, value, identify, and engage with animals and the broader biophysical environment.

John Fraser is a conservation psychologist, architect, and educator with over thirty years working with zoos and aquariums, and studying how they function in society. He is the President and CEO of Knology, a research institute in the United States; serves as Editor-in-Chief of *Curator: The Museum Journal*; and is a past president of the Society for Environment, Population and Conservation Psychology.

Tawnya Switzer is a collaborative writer, who supports thought leaders at Knology, Open Society Foundations, Union Congregational Church, and the Transformative Justice in Education Center at UC Davis. A patterns thinker, Tawnya focuses on dynamics, strategies, and high-impact communications that advance well-being, equity, justice, and sustainability.

THE SOCIAL VALUE OF ZOOS

JOHN FRASER
TAWNYA SWITZER

CAMBRIDGE
UNIVERSITY PRESS

CAMBRIDGE
UNIVERSITY PRESS

University Printing House, Cambridge CB2 8BS, United Kingdom

One Liberty Plaza, 20th Floor, New York, NY 10006, USA

477 Williamstown Road, Port Melbourne, VIC 3207, Australia

314–321, 3rd Floor, Plot 3, Splendor Forum, Jasola District Centre, New Delhi – 110025, India

79 Anson Road, #06–04/06, Singapore 079906

Cambridge University Press is part of the University of Cambridge.

It furthers the University's mission by disseminating knowledge in the pursuit of education, learning, and research at the highest international levels of excellence.

www.cambridge.org
Information on this title: www.cambridge.org/9781108486132
DOI: 10.1017/9781108644679

First published 2021

A catalogue record for this publication is available from the British Library.

Library of Congress Cataloging-in-Publication Data
NAMES: Fraser, John Robert, 1960- author. | Switzer, Tawnya, 1977- author.
TITLE: The social value of zoos / John Fraser, New Knowledge Organization Ltd, Tawnya Switzer.
DESCRIPTION: Cambridge, UK ; New York, NY : Cambridge University Press, 2021. | Includes bibliographical references and index.
IDENTIFIERS: LCCN 2020036363 (print) | LCCN 2020036364 (ebook) | ISBN 9781108486132 (hardback) | ISBN 9781108731812 (paperback) | ISBN 9781108644679 (epub)
SUBJECTS: LCSH: Zoos–Social aspects.
CLASSIFICATION: LCC QL76 .F73 2021 (print) | LCC QL76 (ebook) | DDC 590–dc23
LC record available at https://lccn.loc.gov/2020036363
LC ebook record available at https://lccn.loc.gov/2020036364

ISBN 978-1-108-48613-2 Hardback
ISBN 978-1-108-73181-2 Paperback

Contents

List of Figures   *page* viii
List of Tables   ix
Preface   xi
Acknowledgments   xvi

1   Context   1
    Using Animals to Educate the Masses   3
    Fairfield Osborn: The Zoo as a Vector of Social Change   6
    Reconsidering Mainstream Premises about Learning in Zoo Settings   10
    Chapter Roadmap   13

2   Ontology: Animal Exhibits and Conservation Goals   18
    Developing a Suitable Conservation Agenda Framework   19
    Pursuing a Conservation Education Agenda in Zoos   21
    Why *Do* Zoos and Aquariums Matter? Situating the Role of the Zoo   23
    Implications   27

3   Learning: Social Experiences and Captive Animals   29
    Layers of the Learning Process: Motivation, Meaning, and Morality   31
    Connections to Exhibit Theory and Design   35
       Interpretive Tools   39
    Programming Implications   40

4   Morality: Zoos As Moral Actors   44
    Social Discourse: Evolving Narratives about the Civic Role of Zoos   46
       Zoo-As-Chaos Metaphor   47
    The Ethics of Wild Animal Display   49
       Utilitarian or Rights-Based Philosophical Critiques   51
       Science-Based Critiques of Captive Animal Care and Well-Being   52
       Authenticity Critiques   56
       Moral Development Critiques   59
       Education Benefits: Paths to Deeper Legitimacy   60

5 Pleasure: The Educational Leisure Value Proposition   63
   Positive Psychology and the Pursuit of Meaningful Experiences
     in the Zoo   65
   Understanding the Zoo as an Educational Leisure Venue   71
     Fun and Meaning in the Zoo: Education Psychology Research   73
   Emotional Response Patterns and Educational Leisure Learning
     in Zoo Settings   76
     Seeking and Play   78
     Lust   79
     Fear and Panic   80
     Learning *Is* Pleasurable   83

6 Meaning: Constructing Knowledge through Discourse,
   Dialogue, and Metaphor   84
   Discourse   86
     Discourse Analysis   87
   Dialogue   88
     The Question House Model   90
   Metaphor   91
   Leveraging Discursive Mechanisms for Learning in the Zoo   93

7 Bonding: A Sociobiological Human Need with Important
   Zoo Mission Implications   96
   Human Bonding Pursuits and Outcomes at the Zoo   97
     Family Bonding   98
     School Group Bonding   101
     Affinity Group Bonding   103
     Bonding among Staff and Volunteers   103
   Different Folks, Different Strokes   106

8 Connectedness: Animals, Continuity, and Belonging   107
   Human Connections to Animals and Nature   108
     A Short Summary of the Biophilia Hypothesis   108
     Attitude and Worldview Analysis: Affiliation with Animals 113
   Social–Emotional Implications of Emotional Connections
     with Zoo Animals   116
     Belongings   118
   Leveraging Concern as a Point of Connection   119
     The Value of Connections between Zoo Animals and Staff
       Who Care for Them   124
   From Connection and Concern to a Shifting Moral Compass   127

9   Identity: Discovering Self   131
    Identity Work   133
    Zoos: Unique Sites for "Self"-Learning   136
        Place Identity   138
    Human Responsibility: Environmental Identity Development   141
        Zoos As Sites Where Religious and Environmental Identities
            Overlap   142
    Identity Development Implications   144

10 Activation: Pro-environmental Behavior   146
    Conservation Values and Identities, Precursors to Conservation
        Activism?   147
    Zoos As Activation Entities   151
    Implications   155

11 Impact: Collective Conservation Action   159
    Identity and Society   161
    Collective Identity   162
    Environmental Identity, Developed Collectively at the Zoo   163
    Wildlife Value Orientations   166
    Potential of the Tiny Public   168
    Coordinating Zoo Mission Actors and Action   172
        Communities of Practice   174
        The Social Radiation Approach   176

12 Integration: The Socially Valuable Zoo   180

References   187
Index   207

Figures

10.1 Theory of planned behavior   *page* 149
10.2 Value-belief-norm theory   150
10.3 Psychosocial mechanisms that allow for selective
     disengagement from detrimental conduct in the moral
     control process   153
11.1 Cluster analysis output illustrating how research participants
     associate types of STEM learning sites   166
11.2 Public perception of the degree to which members of the
     public associate frequency of encounters with STEM
     topics with different venue types   168
11.3 Degree to which members of the public associate
     frequency of encounters with a type of learning experience
     and place   169
11.4 Degree to which members of the public associate
     frequency of encounters with a learning topic and place   169
11.5 Collective identity – value-belief-norm theory   171
11.6 Variation in public engagement in personal and civic
     environmental behaviors and frequency of reported
     visitation to zoos and aquariums   178
12.1 Average distribution of adult-only groups,
     adult-with-children groups, and individuals entering
     US zoos and aquariums   182

Tables

4.1 Distinct topic areas that impact public trust in zoos
and aquariums   *page* 49
5.1 Martin Seligman's summary of Christopher Peterson's list
of universal character virtues and strengths   67
8.1 A typology of basic value orientations toward wildlife
and biodiversity   111

Preface

Public debate regarding the value of zoos to society has been lively and ongoing for as long as zoological gardens, in their various historical forms, have been accessible to public audiences. These are inevitably layered and contested conversations, as sociocultural framings of animals and animal captivity are highly variable and fluid, and at no point have all zoos adhered to uniform standards of purpose and animal care.

In the late 1960s, as concern was rising about animal welfare at zoos and aquariums, the special interest group of zoological facilities of the American Association of Museums founded the American Association of Zoological Parks and Aquariums (now known as the Association of Zoos and Aquariums or by its acronym, AZA). That independence coincided with sustained internal and external pressure in developed nations for the world's zoo associations to standardize, legitimize, and police themselves since the late 1970s.

As public cultural institutions, any zoo can be subject to, solicit, and guide public opinion, "the public" being understood as a diverse and intergenerational spectrum of a given society. Public audiences, as conceptualized in this book, might thus include zoogoers and non-zoogoers of all ages; animal rights activists and the volunteers and paid staff who work in zoos; and scientists, religious leaders, and laypeople alike. It was this rich tapestry of perspective and potential that led recently deceased critical museum scholar David Carr to describe public cultural institutions such as zoos as spaces where knowledge, human differences, public policies, and common ideas can be explored with no rush toward resolution.

Today, fewer than 10 percent of the facilities licensed by the US Department of Agriculture to display animals in enclosures for public learning are AZA accredited, meaning that the facility participates in ongoing self-regulation through processes of peer review that confirm significantly more rigorous industry standards. When this manuscript went to press in April 2020, the total number of zoos and aquariums in the United States accredited by the AZA was 240, according to the association's website. While the distinctions between this self-regulated cohort of zoos and aquariums and other facilities that display live animals are very interesting and tend not to be readily understood by the public at large, we focus only on accredited zoos, within and beyond the United States, as the great majority of larger facilities with high attendance around the world are accredited – and thus face the future with shared mission goals and regulatory starting points.

Cross-cultural historical analysis has shown that zoo displays reflect cultural dominion of humans over nature, clearly mapping and reinforcing social narratives about "what is wild" in ways that force visitors to reflect on how humans conceptualize and interact with nature, wild species, and wild spaces.

The world's major zoos have also contributed substantially to what public audiences and experts know about wildlife today, animals' biology and needs, and the threats to survival that a very wide range of species face. Yet, all public cultural institutions must adapt to changing realities in order to maintain their status of socially relevant spaces over time. Museum scholar Steve Weil has suggested that such an undertaking requires ongoing (re)assessment of the extent to which an institution's human and facility capacities are being effectively and efficiently leveraged to achieve mission-related outcomes that are valued by the public.

Civil rights and Holocaust museums, for example, share conscience-related mission goals related to educating and reminding visitors that all people have moral and relational responsibilities to value and protect marginalized human groups. In a similar vein, author John Fraser and his colleague Dan Wharton have long argued that zoos can (and should) be cultural institutions that individually and collectively present a compelling and thoroughly convincing case for why visitors have moral and relational responsibilities to value and protect the natural world.

By the 1990s, zoo decision-makers had become acutely aware that they needed to reposition as the form of cultural institution that would focus on advancing a conservation mission. Seeking to self-define as museums of environmental conscience, accredited zoos within and beyond the United States adopted a formal commitment to conservation before the turn of the millennium – and making that public claim had become central to their marketing and interpretive programs. Concerned about persistent and emerging environmental crises, public audiences in the twenty-first century want to know that "their" zoos are delivering on their conservation promise. Zoos today must thus confirm their operational legitimacy as dynamic cultural spaces advancing a conservation agenda; any zoo that fails to do so faces intense public critique as an outmoded and irrelevant institution.

John Fraser has been exploring the promise of zoos for thirty-five years. Before his eight years with the Wildlife Conservation Society as Director of Interpretive Programs for the New York Zoos and Aquarium, and Founding Director of Public Research and Evaluation, John developed and studied exhibitions for the Toronto and Oregon Zoos. He has impacted zoos around the world as a master planner and is one of the founding social scientists behind the National Network for Ocean and Climate Change Interpretation (NNOCCI), where he remains an active member of the community of practice.

Writer Tawnya Switzer often works with John on complex projects at Knology, a think tank intentional about using transdisciplinary dialogue and cross-sector expertise as starting points for practical social science that explores how inclusive, informed, and cooperative societies can thrive with the natural systems on which we all depend. Working together to better understand and communicate various conditions that appear to spark learning, identity development, and action in informal settings, the authors ultimately articulated a narrative arc that became this book. Tawnya approached these conversations with long-standing discomfort with zoos, but also with absolute certainty that social–emotional connections and reflection on the power and interdependencies of nature's creatures can impact how people relate and commit to sustainable norms. John approached the dialogue from the perspective of an insider with decades of research and leadership experiences within extremely diverse zoo and aquarium settings. The tensions inherent in these different starting points led to rich dialogue about aspects of social value that people attribute to and derive from zoos, the phenomenon and implications of zoogoing, and the functional capacities and unique role of these cultural institutions.

Zoos and aquariums, like the diversity of life they present, come in all shapes and sizes, and these settings unavoidably reflect the values and ideologies of the communities in which they are based. Rather than seek to answer what's good, bad, or unresolved about the conditions, goals, and impacts of specific zoo operations, our goal with this book was to explore the conservation relevance of zoo experiences and what people are looking for, doing, and taking away when they spend time in these institutions. We highlight the implications of our recent research, and the research findings of our colleagues, to illustrate how people perceive, describe, draw meaning from, and value zoos, zoo staff, and on-site experiences. All the research introduced in this book was undertaken at zoos and aquariums that voluntarily seek national association peer accreditation, a status that confirms and ensures dedication to constant improvement and high standards of practice.

While this approach limited our scope of analysis to a well-resourced subset of the very wide range of facilities that fall under the broader zoo and aquarium umbrella, we note that accredited institutions capture a disproportionately high share of zoo and aquarium attendance worldwide, and thus have significant potential for impact. We do not delve deeply into the history of the zoo industry, or defend it, per se, focusing instead on the patterns and implications of zoo usership. To shed light on the latent potential of the zoo form, we draw from a variety of theoretical frameworks and published studies to begin to situate the role of the zoo in the public imagination and draw attention to intersections of social value and conservation mission.

As our manuscript went to press, societies around the world were being brought to a standstill by the first pandemic ever sparked by a coronavirus. Coronavirus is a zoonotic disease that, in 2020, brought into clearer focus the importance of shared understandings, policies, and norms regarding human engagement with wildlife. Too often downplayed as frivolous pleasure and leisure destinations, zoos and aquariums have been important sites for the scientific study of zoonotic diseases, animal and species well-being, and human attitudes to wildlife. So, even as record numbers of individuals sought out live and prerecorded video footage of zoo and aquarium animals while confined to their homes, zoo researchers were being called on as data-informed leaders crucial to the race to understand and combat COVID-19.

This book is our effort to capture what we have come to know from our own work and the work of other scholars – some of whom believe in the form's potential and some who do not. We seek to establish a more nuanced picture of how the narratives, perspectives, and understandings of people who work in zoos, as well as members of the public who do and do not visit zoos, are created, affirmed, and contested within and beyond the communities in which zoos are located. While many of the studies highlighted in the book were conducted in the United States, the data we reference have implications for zoo work, zoo workers, and zoo audiences around the world. It is our hope that this layered tapestry might inform more robust thinking around how zoos can better leverage their power as conservation-focused entities perceived to be credible and trustworthy to ensure that learners in and beyond these settings begin to approach sustainability as a shared concern and establish or affirm conservation activism as an urgent moral priority.

Acknowledgments

This book reflects the incredible generosity of thousands of volunteers, visitors, and zoo professionals who have shared their beliefs and thoughts over the years. The advancement of empirical research norms in zoos and aquariums has not lacked controversy, and we've come across intransigent anti-zoo objectors who reject all such research based on the venue rather than the method or the findings, as well as zoo professionals who created obstacles for fear that findings might fault their work. On the whole, though, most everyone was willing to let the data speak.

In particular, we acknowledge the late Richard Lattis of the Wildlife Conservation Society, who embraced facts and evidence as an excellent basis for improvement, and also the late Dr. Carol Saunders, who was a foundational leader in the field of conservation psychology, an early thought partner who encouraged us to write this book.

We recognize and thank Jackie Ogden, Cynthia Vernon, Kathy Wagner, Joe E. Heimlich, John Falk, Lynn Dierking, Jessica Sickler, Erin Johnson, Martin Storksdieck, and Kelly Reidinger for their important roles advancing the Why Zoos and Aquariums Matter studies. We also thank the AZA's research and technology committee for their support over the years, including our champions Louise Bradshaw, Amy Niedbalski, and Vicki Searles, who have always pushed us to be more specific.

We appreciate the support of our partners in the National Network for Ocean and Climate Change Interpretation (NNOCCI), including our colleagues Billy Spitzer, Hannah Pickard, John Anderson, Nancy Hotchkiss, Nette Pletcher, and Julie Sweetland for their active role shaping our research. We thank professors Janet Swim, Susan Clayton, and Nathan Geiger, who coauthored so many of the studies we report here. The support of the AZA leadership has been unwavering, and we particularly thank Rob Vernon, Amy Rutherford, Nette Pletcher, and Paul Boyle for encouraging us to go further with this work.

Tawnya will be forever grateful that her husband, Greg; sons, Marcel and Pierce; and stepdaughter, Tatiana, adapted to countless nights and weekends when she was completely consumed by this project – and managed to thrive anyway. John gives heartfelt thanks to his husband, Pat, who is endlessly fascinated with giraffes and always agrees to visit one more zoo or aquarium, no matter how small or obscure. And, of course, we deeply value colleagues at Knology, the Institute for Learning Innovation, and the Wildlife Conservation Society whose brilliance never fails to help us refine and better articulate our own thinking – thank you!

This material is based on work supported in part by the National Science Foundation (grant numbers DRL-1713428, DRL-1612729, DRL-1612699, DRL-1240641, DRL-1115217, DUE-12-39775, DUE-1043405, ISE-08-40160); Institute of Museum and Library Services (grant numbers LG-95-17-0058-17, LG-55-14-0148-14, LG-30-08-0035-08, LG-30-03-0255-03, LG 25-05-0102-05, MA-06-12-0143-12, MG-70-18-0009-18); National Oceanic and Atmospheric Administration (grant numbers NA17SEC080001, NA-13-SEC-0080010, NA-10-SEC-0080029); National Endowment for the Humanities (grant BK-50017-06); Environmental Protection Agency (grant NE00A00338); and a variety of private foundations and individuals. The opinions, findings, conclusions, and recommendations in this publication are those of the authors and do not necessarily reflect the views of their funders.

CHAPTER I

Context

Simultaneously social and highly sensory, zoos and aquariums (henceforth referred to simply as "zoos," in most instances) are extremely popular settings that afford notably diverse visitor audiences across the globe with unique informal learning experiences. With more than 700 million visits made to zoos and aquariums that belong to national or regional associations around the world every year, these accredited institutions are fascinating and accessible sites for studying the individual and group-level behaviors, needs, and priorities of animals *and* people.

The World Zoo and Aquarium Association (WAZA), a global alliance of nearly 400 institutions, regional associations, and national federations, gradually adopted a collective conservation mission over the past decades. Given the range of sizes, resources, ethical positions, strengths, and goals of these entitics, though, and the many forms and functions of zoo conservation projects, analysis of mission success continues to be a complicated undertaking.

Spurred by the efforts of the European Association of Zoos and Aquariums (EAZA) and the Association of Zoos and Aquariums (AZA) in the United States, in 2005, WAZA published a conservation strategy document that spelled out the major goal of member institutions as the integration of conservation activities into all aspects of their work. A decade later, the 2015 iteration of this document outlined the following specific steps to integrate a conservation focus into the mission statement and strategic plan of member zoos and aquariums: declare wildlife conservation as a higher purpose; pledge resources to this effort; and create a formal plan to create "a culture of conservation in your staff, communities, governing authorities and donors that gives everyone the opportunity to make a measurable difference."[1]

---

[1] WAZA 2015, 13.

"Conservation" is a fluid term and a pursuit that takes many forms;[2] we define it here as the persistence of wild species and wild spaces. We note that conceptions of "wildness" are similarly fluid and wide-ranging cultural categories – and that both conservation and wildness are being presented, understood, and pursued in new ways in the twenty-first century.[3] Also, to simplify the reading of this book, we refer to nonhuman animals as "animals," though we are fully aware that humans, too, fall within the animal kingdom and appear to have emotional capacities that are indistinguishable from other mammals.

Because most zoo animals cannot ever be viably returned to the wild, given their experiences of domesticity and the limited availability of resources and flourishing wild habitats,[4] some researchers have suggested that the society-level value of any zoo claiming to be a conservation actor must be determined by that institution's impact in any (or all) of the following focus areas:

- species and/or habitat restoration and preservation
- research and monitoring that can inform sound policies about the management of spaces and/or species
- work as trainers enhancing the capacity and skills of those with opportunities to manage species or habitats, impact legislation, or influence existing or future practice
- work as motivational educators promoting conservation knowledge, attitudes, values, and behaviors[5]

Each of these complementary strategies have great potential to shift normative practices in ways that result in a significantly less damaging human footprint on the natural world in future decades, yet there have been very few attempts to measure the long-term consequences of zoos' conservation projects, particularly with respect to education, the final area

---

[2] Nash 1982.

[3] Following the precedent of Myra Shackley 1996, we refer to fauna only, rather than to the more broadly encompassing faunal *and* floral components of a natural environment, when we use the term "wildlife" in this book. Irus Braverman explores the tensions between rhetoric and practice regarding nature and conservation in her 2015 book, *Wild Life*.

[4] For a critique of the ethics and implications of zoos' messaging, goals, and practices regarding reintroducing animals to wild habitats, see Hancocks 1995 and Braverman 2015.

[5] Mace et al. 2007. Whereas Mace et al. situate mission-based zoo training programs as distinct from mission-based zoo education programs, we conceptualize education as sometimes encompassing the development of collective action skills consistent with WAZA's stated conservation education and activism goals. We also note that Rabb 2004 had previously highlighted the importance of measuring impact, presenting a similar list of strategies he felt zoos needed to focus on to become socially relevant and impactful conservation centers.

of focus.[6] Indeed, there is a notable dearth of studies that even attempt to quantify whether zoo visits change the knowledge, attitudes, or behaviors of public audiences, or whether such impacts appear to be consistent or vary across zoos.[7]

What has been established, however, is that the provision of scientific and natural history information about animals and habitats – a default education approach that is still the education strategy most frequently used in zoo settings – has produced no measurable culture change in any cultural context.[8] Certainly, no published research to date shows that zoo visitors have been successfully transformed into a collective movement making measurable impact on biodiversity conservation in the twenty-first century.

A journalistic inquiry into what is currently known about zoos as sociocultural intersections of mission, public perception, visitor experiences, learning, and conservation outcomes, this book draws on conservation psychology and social science research to explore how zoos might develop and deliver more effective learning experiences that promote conservation values and collective action. Recognizing the lack of consensus within and across cultures regarding appropriate human–nature relationships, we look specifically at how meaning-making and informal learning occur in zoo settings; how these cultural institutions contribute to society and measure mission success; and how visitors' priorities and perceptions can inform and converge with new visions for establishing and maintaining sustainable relationships with the world's remaining biodiversity.

Using Animals to Educate the Masses

People love to observe animals. Based on the premise that observing animals might enlighten the general public regarding the wonders of nature, a rapid expansion of the modern zoo movement occurred

---

[6] Mace et al. 2007 point out that measuring and comparing outcomes is notably more complex for public education conservation projects than for research, training, species-focused, and habitat-focused conservation projects. These authors challenge zoos to evaluate success by looking at the longer-term conservation status of wild species or their habitats impacted by conservation projects, and as early as 1995, scholars such as Maple et al. began calling for published studies evaluating the public impact of zoo education programming.

[7] See Balmford et al. 2007 for a review and critique of educational impact assessments. Zimmerman et al.'s 2007 edited volume sought to introduce a framework for measuring the conservation value of zoo-supported projects, including zoos' use of public education to change human behavior.

[8] Fraser and Sickler 2009.

throughout Europe and America after the menagerie connected to the
Jardin des Plantes in Paris, and subsequently the Zoological Gardens of
London, disrupted hundreds of years of upper-class tradition by welcom-
ing the general public in 1793 and 1847, respectively.[9] While the zoological garden was a distinctly European invention, colonizers brought it to
other parts of the world, though in many cases the founding dates and
initial operating norms of zoos outside Europe, North America, Australia,
and New Zealand have been hard to confirm.[10]

In her history of zoos in the United States, historian of science Elizabeth
Hanson notes that by the turn of the twentieth century, more than
100 zoos nationwide were "aiming for the cultural status of scientific
institutions" by offering natural history and general science and nature
knowledge to the masses, and particularly to public school groups.[11]
Operating through public funds and private philanthropy, zoos in the
United States began to be presented and understood as cultural institutions that solidified patriotism and offered education and recreation to the
non-elite. The aim of educating the public and strengthening democracy
in zoo settings also became extremely popular in and beyond Europe
between the First and Second World Wars.

In the 1950s, general optimism around the idea of zoos as democratic
education and recreation settings began to be replaced by deeper understandings of animals and their needs. Threads of opposition against animal
confinement emerged more loudly and gained momentum until critique
became constant in the 1970s.[12] From the 1960s through the 1980s, zoos
faced intense external and internal pressure to respond to challenges

[9] Public audiences in London, though, were only permitted on weekdays; restrictions on public
     access were, in fact, not fully eliminated until 1940. See Baratay and Hardouin-Fugier (2004) for
     details about some of the tensions that arose around efforts to support public access to these sites in
     the interest of public education.
[10] Mullan and Marvin 1987.
[11] Hanson 2002. Mullan and Marvin 1987, Shackley 1996, and Kisling 2000, among others, have also
     provided comprehensive histories detailing how the purpose, role, operations, and design of zoos
     have evolved over time. Historical accounts of zoos tend to begin around 1400 BC, when royal
     menageries and traveling circuses began providing displays of living animals to select groups of
     upper-class citizens or spectacle attendees. Though the first clear case of a menagerie being used for
     scientific purposes was the collection of animals Alexander the Great assembled for his tutor,
     Aristotle, who used these animals as a source of information for his writings on natural history, it
     was not until the end of the seventeenth century and beginning of the eighteenth century that
     several menageries around the world were formally used for an array of science work, according to
     Mullan and Marvin.
[12] Baratay and Hardouin-Fugier 2004. In the early 1970s, activism around smog and water
     contamination reached a tipping point – and the rights to clean water and air became enshrined
     in law in the United States.

regarding the ethical legitimacy of animal confinement[13] and to situate their facilities and mission within the emerging conservation movement.[14] Repositioning zoos as sites for recreation, education, research, *and conservation* was also an adaptive response to burgeoning social concern about an imminent biodiversity crisis[15] and the ethics of captive breeding, which had become a necessary endeavor after the 1973 Endangered Species Act in the United States, limited zoos' ability to collect exotic animals from their native habitats.[16]

In 1992, journalist Alexander Wilson pointed out that by the early 1990s the widespread destruction and loss of habitats and species around the world had shifted the cultural currency of rare and endangered animals; their exoticism, he suggests, had come to be widely perceived as "an exoticism of imminent loss."[17] Indeed, by the mid-1990s, accredited zoos worldwide had formally adopted a new, aligned mission focus that sought to clarify and (re)legitimize the social purpose of these institutions: the conservation of biodiversity.

To support their associations' new claim that accredited zoos are, collectively, a "leading partner" in biodiversity conservation efforts,[18] member institutions began directing a much larger overall volume of human and financial resources to specific conservation-related goals and activities *beyond* sustaining and displaying their living collections. Though most accredited zoos contribute funding and/or expertise to field conservation efforts, public education remains the overwhelming focus of nearly all zoo programming,[19] despite limited evidence to support the claim that zoos help conserve biodiversity in the wild by educating the public.

While decades of international research confirm that substantive learning outcomes do accrue for individuals in zoo settings, evaluation related to the success of conservation education endeavors will require details about the extent to which these institutions are individually and collectively influencing the conservation knowledge, attitudes, and behaviors of public audiences and decision-makers. It remains to be seen whether learning experiences in zoos can and do connect visitors as change-makers

[13] Norton et al. 1995.   [14] Wilson 1992, 247.   [15] Conway 1994.   [16] Hanson 2002.
[17] 247.
[18] This wording came from a 2006 WAZA publication entitled *Zoos and Aquariums of the World*. Rabb 2004 has written about the reasons for new resource allocation directions during that period.
[19] According to their website, AZA-accredited zoos directed $220 million toward mission–supporting field biology conservation endeavors that had a direct impact on animals and habitats in the wild in 2018. Good starting points for description and assessment of what zoos are doing to advance a conservation agenda through in situ conservation, research, reintroductions, and reproductive and medical interventions include waza.org and Zimmerman et al. 2007.

actively seeking local, regional, and/or global-level norms and policies that facilitate the protection of specific forms and systems of life – the outcome we might expect if the "largest potential conservation network globally" were to begin to realize its potential. There is reason for optimism, though, as there is precedent for the zoo as a focused cultural change agent with society-wide impact.

Fairfield Osborn: The Zoo as a Vector of Social Change

Psychologists have shown that environmental values are both malleable[20] and subject to larger cultural frameworks that limit the acceptance and uptake of new environmental ideas.[21] Yet little scholarly attention has been paid to the extent to which cultural institutions contribute to shifts in the social consciousness of the general public regarding the scope of human-kind's responsibilities to the natural environment.[22] In the United States, for example, comprehensive histories of environmental ethics and environmental agendas tend to emphasize the role of academics such as Aldo Leopold, but offer little analysis of how visionary leaders of cultural institutions have used their public platforms to inform public discourse. Such omissions lead to oversight regarding valuable processes that have shaped public consciousness and cultural actors who created the context necessary for academic ideas and/or general acceptance of new moral codes aligned with an environmental ethic.

Born in 1887, Henry Fairfield Osborn Jr. was a fourth-generation descendent of a railway tycoon and son of the president of the American Museum of Natural History who dropped the "Henry" to distinguish himself from his father, according to his daughter, Joan Roth.[23] Both his life and outlook appear to have been characterized by the ongoing negotiation of tragedy and hope. A wealthy Princeton and Cambridge–educated biologist, Fairfield Osborn's writings reveal deep experiences of emotional loss during the First World War, when he served as commander of a Black regiment, and after the war, when his travels throughout Europe served as visual and experiential confirmation of human and ecological tragedy and devastation.

---

[20] Schultz and Tabanico 2007.
[21] Schultz and Zelezny 1999; Oreg and Katz-Gerro 2006; Schultz et al. 2007.
[22] Elizabeth Hanson 2002 has noted that historians in the United States have paid little attention to zoos despite their popularity, perhaps because, among other reasons, they attract diverse, middle-class audiences and do not fit neatly into the categories of analysis typically used by historians.
[23] John Fraser's former colleague, Megan Wells, interviewed Ms. Roth on April 5, 2013.

In 1940, Osborn assumed the presidency of the New York Zoological Society (known today as the Wildlife Conservation Society), and he quickly began using the society's first zoological park in the Bronx (known today as the Bronx Zoo)[24] as a platform to raise nationwide debate regarding the need to curb overconsumption in view of widespread environmental degradation and expanding populations. Osborn's publications and radio broadcasts, and various management pedagogies he instituted at the New York Zoological Society from over three decades of leadership, were related efforts to disseminate transformative ideas about humans' connectedness to natural systems and the rights of nature. These narratives found form through experimental zoo initiatives favorably responded to by Bronx Zoo audiences.

Before the First World War and in a second wave of development in the 1930s, German zoo director Carl Hagenbeck had revolutionized zoo design by designing landscapes that created the experience of viewing African savannah animals as if they were sharing the same habitat. Rather than barred cages, he used a series of moats to prevent animals moving between exhibits, giving visitors the opportunity to look across those moats and perceive animals living together in harmony. While this innovation was little known outside the zoo circles of Western Europe, it captured the attention of Fairfield Osborn during his travels through Europe as a young board member of New York Zoological Society. When he became chair of the board in 1938, he brought this revolutionary idea to the redevelopment planning process for the Bronx Zoo's proposed African Plains exhibit, promoting a New York architect as the visionary behind the new concept in annual reports presenting the design.

The African Plains exhibit opened in 1941, as the United States was on the verge of entering the European war that thoroughly destroyed Hagenbeck's zoo. Unlike Hagenbeck's naturalistic displays, though, African Plains was arguably the first display that intentionally sought to educate visitors about ecological relationships by grouping animals as they might be found in nature, rather than by taxonomy or extremely broadly by continent. Osborn promoted the idea that his exhibition would help audiences think of humans as part of and dependent upon natural systems. The exhibit's popularity confirmed the appeal of presentations of nature that were completely different from previous approaches to zoo design,

---

[24] See Hanson 2002 and Fraser 2004 for in-depth historical context regarding the naming of this park.

which had suggested a highly organizable natural order disconnected from human presence.[25]

Typically cast by historians as both a noble conservationist figure and a scandalous white supremacist, Osborn's predecessor, William T. Hornaday, founding director of the New York Zoological Society's Bronx Park (Zoo), had been a leader who approached change as a process best achieved through engagement with the elite. While both men were deeply committed to advocacy and repositioning zoos as centers for education about wildlife and conservation, they had divergent perspectives about whether the promise and purpose of the zoo are best leveraged through elitist or populist institutional ideals.[26] Allowing zoo visitors to bring their own cameras and supporting field biology research and field conservation initiatives in Africa and Southeast Asia, Osborn had new ideas about how people should share experiences of learning about and from animals.

As the New York Zoological Society began to integrate a social consciousness lens into its education strategy and purpose under Osborn's leadership,[27] he was able to secure support and resources for experiments in environmental education films. Review of his nearly thirty years as board chair show that Osborn directed early funding to support the nascent research of an emerging group of environmentalists that included Aldo Leopold in the 1930s and Rachel Carson in the 1950s. Osborn published wildly popular pamphlets and books about wildlife, drawing on his family connections with the publishing world in New York. Building on the success of his books, he emerged as a national radio personality who sought to develop among audiences an emerging critical consciousness regarding social responsibilities toward nature. These were

---

[25] Baratay and Hardoiun-Fugier 2004 have noted that interest in non-romanticized perspectives of nature grew in lockstep with the burgeoning wildlife protection measures in Europe and North America between 1890 and the early 1900s. Rather than continue to attempt instilling moral values in an illiterate urban public through romanticized encounters with living dioramas, zoo operators around the world started using highly organized taxonomic display strategies.

[26] Hornaday, also a prolific writer, focused heavily on the legislative domain and provided his publications to members of Congress and people of influence, whereas Osborn was notably focused on appealing to the general public to shift social and cultural norms. Unlike Hornaday, John Muir, and other notable influencers who communicated conservation values in an elite milieu, Osborn was effective at delivering his messages about the value of wilderness to extremely diverse audiences; his own publications demonstrate that the notion that wilderness was worthy of protection had become well established in the public consciousness in the 1940s.

[27] Memos from the president to the director requested a history of the New York Zoological Society's conservation work. Margin notes in an early draft indicate a preference for principles of decision-making rather than individual projects. Records of these transactions and activities can be found in the archives of the Wildlife Conservation Society's Bronx Zoo library.

multipronged strategies to establish among diverse audiences a shared commitment to conservation values by contributing to an emotional narrative of impending destruction of natural and human systems.[28]

Some of Osborn's early publications were political efforts to end American soldiers' habit of shooting songbirds when not on duty in Pacific Island countries. In 1944 he edited and published *The Pacific World*, a collection of manuscripts written by science and education professionals encouraging bird-watching and natural history dissemination so those serving overseas would be encouraged to take responsibility for the conservation of biodiversity in the oceans, lands, and skies of the Pacific Ocean.[29] Four years later, *Our Plundered Planet* featured praise from Aldous Huxley and Eleanor Roosevelt on the back cover, went through several printings, and was also published in French and German. Osborn had established himself as a leader capable of impacting public dialogue and critical consciousness.

These publications had a transformative impact on public dialogue and critical consciousness that ascribed nature with having inalienable rights. Using global evidence and a passionate call to action, Osborn sought to arouse public opinion around environmental stewardship as a new ethical attitude – and cooperation between citizens, industry, and government worldwide – based on the inalienable rights of nature. He also claimed that responsibility to future generations was imperative to the survival of nations. When he published *The Limits of the Earth* in 1953, Osborn described cultural contexts, economic conditions, "social and moral points of view," and "spiritual concepts" as core aspects of the fluid relationships between people and natural resources.

Ultimately, Osborn was able to use moral and scientific messaging, through his platform of institutional legitimacy and status as an admired philanthropist, to simultaneously shape public consciousness and public policy. Postwar rhetoric did begin to include ideas about the moral and practical need for the long-term conservation of natural resources. Over time, Osborn wielded enough influence to ensure the protection of significant natural resources in the Pacific Islands during and after the Second World War, and the New York Zoological Society had a publicly visible role in the establishment of Jackson Hole Wildlife Park, which

---

[28] The emotional narrative of impending destruction of natural and human systems presented in Osborn's radio broadcasts and publications has been identified as apocalyptic. See Netzley 1999, who characterizes Osborn's work as an exemplar of the apocalyptic antihuman narrative.

[29] *The Pacific World* (1944) was published in two editions in order to reach Americans serving in the military *and* the general public.

became part of Grand Teton National Park.[30] While falling short of Osborn's commitment to ascribing nature intrinsic value, we note that several of the United Nations' recently established Sustainable Development Goals (SDGs) do build on Osborn's legacy by seeking to capture both tragedy and hope within an operational framework for sustained change, and by pointing to ethical interpretations of human–nature relationships as a starting point for discussion and action.[31]

Biologist David Ehrenfeld has noted that, at the public venue level, dissemination of new ideas can only succeed if the audience is complicit in recasting the narrative because the institution will otherwise lose social relevance, funding, and support. Ehrenfeld also points out that few institutions ever survive shifts in a dominant worldview, noting, furthermore, that the few that do tend to be "changed out of all recognition."[32] We note that Fairfield Osborn, who died in 1969, successfully established a precedent of institutional transformation that repositioned a leading zoo and its representatives as cultural change agents by leveraging the institution's perceived legitimacy and relevance as a civic actor. Of particular relevance for the focal direction of this book, one very interesting outcome of Osborn's multipronged strategies to engage public audiences around the world with new ideas of connectedness, morality, and responsibility was a dramatic rise in Bronx Zoo attendance in the 1950s, particularly by families, a pattern to which we will return in subsequent chapters.

Reconsidering Mainstream Premises about Learning in Zoo Settings

Even as zoos have continued to evolve in their pursuit of conservation as a collective social purpose, researchers have had limited ability to predict or confirm the achievement of conservation-related education outcomes for several reasons. We note that while there has been significant philosophical reasoning around zoos' animal care practices in the past several decades, there continues to be scant empirical research into how zoos are advancing

---

[30] Records of these transactions and activities can be found in the archives of the Wildlife Conservation Society's Bronx Zoo library.

[31] The SDGs define a future development agenda that clearly situates the ecological, social, and economic spheres as interrelated domains that have direct and important impacts on human rights, justice, health, and well-being. For analysis of the ethical structure of the SDGs, and associated interpretations and ideas about implications for how to value and make decisions about nature, see Keitsch 2018.

[32] Ehrenfeld 1995, xvii.

community or societal needs, conceptual thinking, or community dialogue related to conservation. Furthermore, the default pedagogy that frames how learning is approached in many zoos presumes that exposure to science-related data points and animal observation motivate visitors to act and are an efficient path to mission fulfillment.

Researchers Joe E. Heimlich and John H. Falk describe the learning process as "taking in, organizing, and making meaning of the world," a lifelong, natural process that "happens consciously, subconsciously, and often only coincidentally with the intentions of the individual."[33] This is the definition we will use for the purposes of this book, while noting that learning can include changes in cognition, affect, attitudes, and/or behavior.[34] In a similar vein, we understand "environmental education" to be the process of "recognizing values and clarifying concepts in order to develop skills and attitudes necessary to understand and appreciate the interrelatedness among humans, culture, and the biophysical surroundings."[35]

Yet, other than an emerging body of conservation psychology scholarship and a few recent social science assessments that will be highlighted throughout the book, zoo research to date has overwhelmingly measured learning only with respect to the acquisition of factual science and natural history narrative knowledge. Furthermore, almost all zoo research has relied on individual-level visitor assessment, though their shared mission clearly situates accredited zoos as social sphere entities and actors.[36]

We aim to counter these weaknesses and clarify opportunities for mission success by instead exploring the range of visitor experiences and free-choice learning that occurs in the social settings of zoos – with attention to self-discovery, shared learning, and the role of motivation and repeat visitation in the development of conservation knowledge, attitudes, and behavior. Social science researchers in the twenty-first century have been accumulating findings that have important implications for how zoos approach their mission of conservation through education.

Beyond identifying zoos as sites where learning occurs, we will introduce data suggesting that public audiences tend to see zoos as trusted vectors of social change; expect to learn about nature and conservation in zoo settings; and hope their experience will include moral development

---

[33] Heimlich and Falk 2009, 11.
[34] For an overview of this conceptualization of learning, see Falk et al. 2009, 6.
[35] This is the language of the International Union for Conservation of Nature and Natural Resources (1971).
[36] Fraser and Sickler 2007.

and identity exploration. We also highlight emerging evidence that experiences in zoo settings can galvanize visitors toward action around specific conservation topics. These findings affirm that, in addition to leveraging their living collections as a powerful educational tool, zoos can and should experiment with how best to leverage their cultural capital as perceived agents of moral change to advance an ambitious conservation agenda. Furthermore, the range of professionals and volunteers who impact visitor experiences in these institutions should be understood and supported as valuable actors in a mission-grounded movement to advance conservation-based policies and social norms.

As sites in the public domain in which civic responsibilities and values are explicit topics of discourse and negotiation, museums are granted unique authority by the communities that support them to offer themes, exhibits, and exhibitions that provoke reflection and narrative development related to personal and communal values, roles, and priorities within a connected biosphere.[37] Former WAZA President Lee Ehmke pointed out in 2015 that "failure to act more forcefully and effectively will threaten the business model and social license that allow our institutions to exist and thrive."[38] We agree, and set out to identify and analyze recent research on zoo visitor experiences that might shed light on how audiences today situate zoos as part of their learning world and how public perceptions appear to qualify or limit zoos as conservation communicators – patterns that point to the capacity of this type of institution to fulfill its mission of awakening and affirming widespread consciousness and action around biodiversity and ecological integrity.

We believe this is the first applications book that integrates a conservation psychology framework into a larger concept of social change and uses user-group perspectives to clarify how social science research can generate practical knowledge that strengthens zoos' capacities for meeting conservation mission goals through visitor engagement. By beginning to identify the conditions in which learning in and from zoos and aquariums appears to impact visitors' conservation-related awareness, attitudes, and behaviors, we seek to lay some necessary groundwork for the development of a robust body of future research around teaching and learning practices that advance conservation goals by tangibly impacting personal and collective commitments to conservation outcomes. Like Fairfield Osborn, we seek to show the breadth of perspectives around conservation issues as well as potential for identifying common ground and feel a conservation agenda is

[37] Fraser 2004.   [38] WAZA 2015, 9.

best advanced through cooperative study, institutional leadership, and activism based on shared morality.[39]

Chapter Roadmap

Throughout this book, we introduce recent scholarship from various domains of social science research that may help zoos better leverage their human and facility capacities to achieve mission-related outcomes valued by the public. Chapter 2 delves more deeply into the evolution of zoos and aquariums in the conservation movement, highlighting conservation biology and conservation psychology as frameworks that emerged in part from zoo industry agendas and research. While default education programming and evaluation tools in most zoos continue to be based on an information transfer model of learning, we introduce research that suggests this approach does not speak to public audiences in the language of their values and expectations. Regarding how these findings might inform new questions, approaches, and measurements of zoo mission success, we suggest that an array of strategies to promote and support the moral and ethical engagement of individuals, groups, communities, and societies to protect and preserve wild species and wild spaces must complement educational programming designed to increase factual knowledge. We argue that zoo messaging and experiences must help learners of all ages identify themselves as embedded within natural systems – and see and value biodiversity as valuable, irreplaceable, and worthy of active protection and behavior change.

To situate learners within the unique context of zoos and aquariums, and the zoo as a conservation learning tool, Chapter 3 considers these institutions from the perspective of those it serves. Zoogoers' perceptions, motivations, priorities, and preexisting conservation-related knowledge, concerns, and experiences are integral parts of the learning experience. Researchers have confirmed that zoo visitors build on, develop, and reinforce layered ideas and narratives they find personally meaningful when they establish or negotiate connections and coherence during their on-site experiences. This helps explain why zoogoers' descriptions of their zoo experiences tend to seem completely disconnected from the logico-

---

[39] Osborn, as editor of a collection of essays entitled *Our Crowded Planet* published in 1962, gave contributing authors free reign to focus on any aspect of the relationship between population and resources they wished to explore, in order to get optimal thinking and find points of convergence across domains as varied as philosophy, religion, biology, and zoology.

deductive science learning pedagogy zoos have historically prioritized. In this chapter, we consider the many ways learning has been witnessed in zoos and why the presence of living animals may produce learning pathways unrealizable in other settings. To make these connections, we begin to unpack what appears to be happening in the minds of diverse users of the zoo, an approach that allows us to describe multiple opportunities for learning and various stories that are present. We conclude that zoogoers' emergent, on-site experiences and motivations for visiting do often converge with zoos' conservation education goals.

While rumblings about zoos as chaotic incarceration have persisted since the London Zoo was first made accessible to the general public, sustained challenges regarding the rights and inherent responsibilities of zoos as moral actors have been a persistent feature of zoo operations for the past fifty years. Chapter 4 delves into the context, understandings, and assumptions that seem to ground common narratives of zoos as symbols, destinations, and public voices with authority in civil discourse. Recognizing that metaphoric and moral critiques have long challenged the legitimacy of zoos as moral actors and leaders in the environmental movement, we look at how philosophers, public audiences, zoo and animal behavior professionals, and critics perceive, present, value, and contest the civic role of zoos in contemporary society. We deconstruct common critiques, introducing research relevant to narratives about the legitimacy of these sites as educational venues and beginning to outline why we feel zoo staff's knowledge of and care for animals within and beyond zoo settings should be a key component of strategies to advance conservation mission goals.

To understand and categorize their visitors, zoos and aquariums have historically focused on demographic labels such as age, ethnicity, social group, level of formal education, and frequency of visitation, or marketing categories such as "educational" visitor versus "leisure" visitor. Chapter 5 takes a layered look at the visiting phenomenon, noting that while natural history museums, art galleries, and zoological gardens all feature collections curated for presentation and public interpretation, the former settings are typically considered high-culture destinations for learning, whereas zoos and aquariums have no such status and are often trivialized as contexts for recreation and amusement. Building from psychological understandings of positive psychology and various conditions that provoke pleasure, fun, meaning-making, and emotional response experiences, we explore the distinct value of the zoo as a social place of educational leisure that offers public audiences many points of sensory stimulation and potential engagement due to the proximity of numerous types and taxa of live animals.

Chapter 6 introduces discourse analysis and explores some of the discursive tools that visitors use to inform, organize, and interpret experiences they have at the zoo. We highlight dialogue and metaphor as distinctly relevant discursive mechanisms for the learning that tends to occur as zoogoers negotiate multiple narratives and multisensory, emotional experiences during their visit – and simultaneously and subsequently integrate those experiences into their existing mental, narrative, and moral frameworks. Because nature and human–nature relationships are neither fixed nor uniform, we suggest that, rather than seek to enforce a singular narrative, agenda, and strategy for conservation, zoo leaders should focus on establishing and developing their spaces and their staff as flexible resources equipped to facilitate idea sharing and accountability through civic mechanisms of discourse development and dialogic exchange. Because on-site educators can deepen visitors' engagement and retention of information but are not visitors' sole source of ideas or information, we focus on how zoo staff can use context and metaphor to flexibly initiate or adapt discourse and dialogue to fit the setting and different zoogoers' existing knowledge and intentions. We note that context becomes an important scaffold when spaces for knowledge sharing and reasoning enable diverse learners with varied foundational knowledge and perspectives to reach new levels of understanding, commitment, and conflict resolution.

To fine-tune understandings of the cultural value of zoos today and further explore how these popular institutions can better achieve their collective conservation mission, Chapter 7 is a deep dive into one of the most important and highly valued dimensions of zoo visiting: human bonding. Having established in Chapters 3 and 5 that public audiences overwhelmingly value zoos as settings for (and of) social engagement, and tend to situate social experiences as a core function that differentiates zoos from other museum types, here we explore human bonding as a sociobiological human need – and a strong and consistent motivation and component of zoogoing – noting that on-site bonding is a social capital development process important to zoos' collective conservation mission. We outline the sociological concept and psychological value of human bonding and use observation and evidence to assess how live animal stimuli in the safe setting of the zoo provide rich and layered opportunities for dialogue and shared meaning-making, processes that have been shown to build trust, strengthen social bonds, and contribute to the establishment of shared perceptions about valuing and caring for animals and nature.

Non-social emotional and mental impacts of on-site nature experiences also emerge as a common theme when people describe zoos and zoogoing. Chapter 8 thus delves into what is known about mental models of connectedness, continuity, and belonging that appear to be linked to empathy, guilt, concern, and care – emotional responses that often arise automatically when humans encounter live animals. In addition to a psychological exploration of abstractions of nature and the concept of biophilia, we highlight patterns of affiliation, caring, and connectedness that emerge in zoo settings to shed light on the extent to which experiences that spark empathy toward animals can facilitate the development of moral emotional responses likely to ground or reinforce a conservation ethic. Building on primary research about the implicit connections people tend to develop during zoo visits – irrespective of the fact that zoo spaces are clearly human-designed – we demonstrate that zoo experiences overall impact humans' capacity to connect and extend their scope of care to nonhuman entities in important ways. In particular, we consider why and how emotional relationships between zoo animals and the staff who care for them can and should be at the forefront of efforts to facilitate emotional connections to animals and caregiving that might build on and inform zoo users' existing ideas, concerns, and motivations.

Building from the importance of awe, wonder, and care as points of entry that foster humans' emotional bonds with animals and lead to internal tensions around the moral and relational implications of that connectedness, Chapter 9 explores related processes from an internal psychological perspective. We introduce the idea that, in ways other cultural venues cannot, zoos help individuals understand more about who they are, suggesting that personal identification with animals and conservation goals can predispose people to act. We look specifically at place identity, environmental identity, religious identity, and how identity work might lead zoogoers further toward an integrated ethics of compassion that features feeling for, and with, other beings.

Chapter 10 outlines conditions that appear to support the emergence of conservation values and identities that take the form of action, looking at current understandings of human behavior and what is known about zoos' institutional power and capacities. Returning to the relevance and impact of the zoo as a museum with conservation mission goals, we explore theories and evidence relevant to the potential of zoos as activation entities able to shift zoo users from individual-level processes of values-focused identity work to the active (and hopefully ongoing) pursuit of a conservation agenda.

Emerging research is now demonstrating the importance of recognizing zoos as cultural institutions that offer eudaemonic value, sites in which self-actualization exemplifies personal expressiveness, collective consciousness, vitality, and direction – characteristics and outcomes many zoo users seek as members of a larger community movement for social change to protect wildlife and natural systems. Going beyond narrow ideas of the zoo as a venue for hedonic personal fulfillment, Chapter 11 explores the issue of collective identity theory in the context of zoo experiences, suggesting that zoos have more direct value as catalysts for social change than has been reported in the peer-reviewed literature to date. Here, we focus in a fresh way on the zoo movement itself, exploring how complementary and collective work across the sector can produce society-level change – a far departure from the idea and goal of inspiring an individual, through a single zoo visit, to individual-level actions that hopefully add up to meaningful impact, somewhere, somehow, at some point. We conclude with the thought-provoking example of a community of practice working across a distributed network of zoos, aquariums, and nature centers to deliver coordinated, context-relevant conservation education and messaging to advance a conservation ethic and spark action among potential allies in numerous spheres.

Chapter 12 reviews the latent potential of accredited zoos: unique affordances and the public trust and authority needed to truly advance a conservation agenda by integrating complementary domains as relevant and as wide-ranging as pleasure, meaning, values, ethics, morality, connectedness, emotion, learning, and identity. These factors, coupled with clear pathways and a call to action, can motivate and sustain personal behavior change and collective undertakings to establish sustainable stewardship norms that support biodiversity *and* human well-being.

CHAPTER 2

Ontology
Animal Exhibits and Conservation Goals

Threats to biodiversity occur where ecosystems and human social systems intersect. Biodiversity protection and preservation occur at the same intersections. Now a pursuit at the global level, The United Nations Strategic Plan for Biodiversity 2011–2020 is one example of a formal endeavor currently underway to halt the loss of biodiversity. Target 1 reads, *"By 2020, at the latest, people are aware of the values of biodiversity and the steps they can take to conserve and use it sustainably."* If accredited zoos deliver effective educational programming that helps zoogoers value biodiversity and identify pathways to action each of the 700+ million times the turnstiles are used at these institutions in a given year, Target 1 will be that much easier to meet. Zoo management and educational programming teams thus have a strong interest in understanding what conditions appear to help a person become more aware of why biodiversity is valuable and how they can advance a conservation agenda, as well as, taking this a step further, what conditions appear to help people actively pursue sustainable conservation norms.

The past fifteen years have seen increased recognition that conservation learning in zoo settings must be framed and assessed using social science tools that capture social phenomena linked to evolving attitudes and behaviors related to conservation. Published in 2008, the National Science Foundation's framework for evaluating informal science education (learning that occurs outside the classroom) lists five important impact areas: awareness, knowledge, and understanding; engagement or interest; attitude; behavior; and skills.[1] It stands to reason, then, that frameworks and strategies to pursue and assess zoos' collective conservation mission impact should capture processes and outcomes as varied as learning, decision-making, identity formation, public perception, and cultural change. Yet, a vast majority of environmental education efforts and

[1] Friedman 2008.

assessment of those efforts, within and beyond zoos, have, until surprisingly recently, been very narrowly grounded in ecology, which is the scientific exploration of relationships between living beings and their biophysical environment.[2]

Developing a Suitable Conservation Agenda Framework

Conservation biology was originally defined in the 1980s as a transdisciplinary reframing of the biological sciences that might bring coherence to multidisciplinary research aimed at mitigating biodiversity loss. Recognizing *human* values and behavior as the root of "environmental problems," conservation biology concepts emerged through small population studies at zoos. This framework has as a core tenet the assumption that individual, local, regional, and global decision-makers will rationally follow conservation action guidelines suggested by the science community.[3] Yet, by the dawn of the twenty-first century it had become painfully clear that exposure to science-based conservation information was *not* decreasing the pace of resource consumption or global ecological destruction,[4] and social scientists had established that neither cultural norms[5] nor personal self-interest[6] are adequate conditions for widespread environmentally protective behavior, anyway.

Dr. Carol Saunders, an animal behaviorist working in education studies at the Brookfield Zoo, and the zoo's director, George Rabb, one of the founders of the International Union for the Conservation of Nature, were troubled by this growing awareness that conservation problems were accelerating and there appeared to be little evidence to stem that tide. They questioned how the zoo might become more actively engaged in driving measurable conservation. In 2001, Saunders convened a group of psychologists who had expressed interest in environmental issues to look across their branches of focus for information and concepts potentially relevant to an environmental conservation mission. Participants pointed out that, beyond a perceived overhaul of mainstream lifestyle norms and economic models, conservation-focused behaviors involve *empathy,*

---

[2] Fraser and Brandt 2013 situate and speak to this reality.   [3] Soulé 1986, 1987.
[4] See Mascia et al. 2003 regarding the convergence of concerns from the fields of economics, sociology, environmental history, political science, and psychology. A related outcome has been the emergence of anthrozoology, an interdisciplinary field that includes scholars from fields such as anthropology, sociology, biology, history, and philosophy, and explores interactions between humans and other animals, with particular focus on quantifying the positive effects of human–animal relationships on either party and studying such interactions.
[5] Dietz and Stern 1995.   [6] Stern and Dietz 1994; Schultz 2000.

motivation, decision-making, and the development of new social norms. All are psychological factors.

In 2003, building on Saunders' fresh thinking, the Society for Conservation Biology tasked a cross-disciplinary working group with developing a robust body of conservation-related social science research to complement the conservation biology research that had dominated the literature for over a decade.[7] The goal was to advance understandings related to *how* people develop conservation values, attitudes, and (new) behavior patterns.[8] Together, this small team developed a conservation proposition to parallel conservation biology, introducing in a special issue of *Human Ecology Review* **conservation psychology** as an integrative pedagogic framework for research-based analysis of how individual attitudes, knowledge, and motivations can be developed as behavioral levers that lead, ultimately, to more environmentally protective and sustainable society-level norms.[9] Researchers subsequently began seeking more nuanced understandings of what might motivate changes in society-level environmentally protective values, behaviors, and social norms – a line of focus that led to wider study of social and moral domain concepts such as ideology, values, and ethics.[10]

A note about terms: Whereas *ideology* refers to a broader and more inclusive worldview perspective, *values* are enduring beliefs about desired end states and appropriate standards of conduct – guidelines, essentially – for what to think and do. Critical to the intergenerational transmission of culture and primarily shaped and cumulatively established during childhood, values inform human behavior and facilitate social cohesion through their influence on attitudes and norms.[11] *Ethics* impose restraints on individuals in the interests of the community.

Nearly two decades after Aldo Leopold suggested the elite adopt a "land ethic" attitude and action pattern of moral responsibility and ethical obligation to the biosphere, James Lovelock asserted that the earth should be understood as a complex and synergistic whole that includes all living

---

[7] The Society for Conservation Biology is a global community of conservation scientists and practitioners that has been promoting information exchange and building social science capacities among conservation practitioners for over three decades.

[8] Clayton and Brook 2005.

[9] Saunders 2003. Early conservation psychology work included exploration of the differences among egoistic values, characterized by little sympathy beyond the self or immediate family; altruistic values, characterized by empathy conferred to entities beyond the family; and biospheric values, characterized by prioritization of systemic integrity over the rights of individuals (see, for example, Schultz 2000, 2002; Kaiser and Schultz 2009).

[10] Saunders and Myers 2003.

[11] For an overview of the theoretical underpinnings of these concepts and implications for wildlife management, see Manfredo et al. 2009.

organisms and their inorganic surroundings, proposing that the natural environment thus be accorded ethical rights. Similarly grounded in relational perspectives that are both ancient and widespread in religious history,[12] these authors' claims that the environment has intrinsic value (unrelated to its usefulness to humans) offer a new concept of global citizenship, rather than narrow focus on the protection of specific species.[13]

While psychological research on values and attitudes shows that most humans are motivated by communal concerns,[14] researchers have limited understanding of the factors and narratives that help people "know" that their personal experience and actions are deeply connected to the health of the biosphere on which we all depend. With the short- to medium-term goal of changing behavior among individuals and communities and motivating conservation action locally and globally, conservation psychology seeks to provide the conservation community with information related to three broad questions:

1   What experiences are necessary to ensure that people develop an emotional concern for nature?
2   How can environmental information be most effectively presented so people will accept it?
3   How can people be encouraged to engage in conservation action?[15]

The conservation psychology approach has gained significant traction as a starting point for assessing experiences that draw/push people to place higher value on wild species and wild spaces by evoking empathy and connection in public settings such as art museums and zoos. The remainder of this chapter begins to explore zoos as cultural institutions that may offer unique opportunities for cognitive, emotional, and moral connections and responses to animals and animal well-being that help zoogoers transcend passive, short-term, self-interest.

Pursuing a Conservation Education Agenda in Zoos

Zoos and aquarium management teams were early adopters of the conservation biology agenda, helping found the Society for Conservation

---

[12] See, for example, Kinsley 1995.
[13] For a detailed review of the conceptualizations, claims, and impacts of environmental thinkers in the United States, see Nash 1967. We also note that Manfredo and colleagues have further unpacked the evolution of these narratives with a quadrant categorization strategy that illustrates how society his migrated from a more utilitarian perspective to a mutualist narrative as the country urbanized; see Manfredo et al. 2003.
[14] Fraser and Sickler 2008a.   [15] Saunders and Myers 2003.

Biology, in 1985,[16] and subsequently incorporating conservation biology principles into field science, scholarship, and messaging.[17] Accordingly, exhibit goals and education programming in many of the world's more prominent zoos shifted toward endeavors to advance society-level conservation values.[18]

Conservation education strategies in zoos and museums in the United States have a long history, though, and many such strategies were already in place by the 1980s. Soon after Fairfield Osborn began displaying captive animals in landscape groupings that mimicked natural settings, the Arizona-Sonora Desert Museum used similar design strategies to achieve a similar goal: provoking reflection about the connections between live animals and their environments. These models, in turn, influenced the design of ambitious new zoos in other US cities, such as Milwaukee, Wisconsin and San Diego, California, in the 1950s and 1960s.[19]

William Conway, who succeeded Fairfield Osborn as Director of the New York Zoological Society, published an article that outlined a vision of zoos as natural history and conservation centers. "How to Exhibit a Bullfrog: A Bed-Time Story for Zoo Men" made the case that zoos should provide zoogoers "*a new intellectual reference point, meaningful and aesthetically compelling; a view of another sensory and social world; a feeling of personal interest in diminishing wild creatures and collective responsibility for their future which is so closely linked to that of man.*"[20]

Over time, this vision gained widespread acceptance, and zoo designers and zoo managers alike have been grappling ever since with the potential of captive animal displays to impact visitors' personal and political behaviors. Bob Mullan and Garry Marvin stated in *Zoo Culture* in 1987, "it is this potential of the zoo experience to create conditions for human *self*-definition that makes zoos as institutions such revealing areas of investigation" (emphasis added).[21] People in the zoo, furthermore, certainly offer conservation researchers and advocates what appears to be a culturally representative slice of all levels of society and source of information about cultural norms, perceptions and priorities, and values.

At the turn of the century, most zoo-related research pursuits reflected the perception that zoo learning is more focused on (and more beneficial to) children than adults.[22] This perception had roots in an epistemological

---

[16] Soulé 1987.   [17] WAZA 2005.   [18] Dierking et al. 2002.
[19] See Hanson 2002 for historical perspective.   [20] Conway 1968.   [21] Ibid., xviii.
[22] See Dierking et al.'s 2002 comprehensive literature review of studies that had validated children's science learning in zoo contexts. We note that while several of the reviewed studies demonstrated that zoo exhibits do significantly impact education outcomes, learning was not clearly defined in

view of learning that had long situated the zoo as an institutional authority that transfers science and natural history knowledge to visitors who arrive with limited knowledge through choreographed exposure to intentionally presented information.[23] Research supporting the conclusion that experiences in accredited zoos tangibly impact what adult visitors know and understand about conservation was anecdotal and limited in scope, could not be applied to zoo settings overall, and could not demonstrate shifts in conservation understanding.[24]

Furthermore, even as conservation psychology researchers began to identify factors that appear to impact individuals' adoption of a conservation ethic, and thereby shed light on how zoos might optimize mission-focused educational programming, a glaring knowledge gap remained. If public audiences do not *value* zoos for their conservation education mission, even data-informed educational experiences are highly unlikely to be effective because different types of museum have distinct roles in the eyes of the public – and these perceptions are known to influence what a visitor expects and learns during cultural institution experiences.[25]

Why *Do* Zoos and Aquariums Matter? Situating the Role of the Zoo

Historian of science Elizabeth Hanson (2002) claims that prior to the thrust in marketing research initiated since the turn of the twenty-first century, zoos had been largely ignored by scholars for two key reasons: These spaces have long been characterized by middle-class appeal that falls between the categories of analysis typically used in the social sciences; and the animals themselves seem to occupy a hybrid space that is neither circus or high culture, wild or domestic. At the turn of the millennium, she pointed out that there were few studies conducted on the social value and purpose of zoos, and those that had been completed during the mid-twentieth century tended to offer little beyond demonstrating that these

---

most of these studies and the findings could not be generalized to zoo visits overall. Furthermore, some of the reviewed studies, as well as subsequent research in the United Kingdom (Balmford et al. 2007), illustrated that little cognitive learning gain can be directly attributed to a zoo visit.

[23] Wilson 1992 and Rothfels 2002 have demonstrated that naturalistic zoo design strategies emerged directly from a late nineteenth century romantic view of wild nature within the agenda and context of teaching natural history to zoo visitors.

[24] Dierking et al.'s 2002 analysis of zoo education conference presentations between 1985 and 2000, for example, shows that mention of research or data on the *impact* of conservation education in zoo settings was extremely rare, despite consistent reference to natural history knowledge transfer.

[25] Pekarik et al. 1999; Fraser 2017.

institutions have served primarily as a recreation experience for a widely diverse group of middle-class visitors.[26] Further complicating these matters, at that time market researchers had consistently found that no single motivation or demographic characteristic can be attributed to zoo visitors – or explain the widespread middle-class appeal of zoos.[27]

Hanson offered a fair indictment of the claims to social change that zoos were making at the time, and her summary was also being felt internally at these same institutions. The public accountability movement that emerged in the late 1970s as a conservative social narrative, had gained steady momentum, leading, over time, to a global restructuring or revisioning of public sector contributions to education.[28] The anti-zoo movement had seized on related narratives regarding the public financing of zoos, focusing, in part on claims around educational mission tied to the ethical obligations of animal caregivers. Zoo staff found themselves having to devote a great deal more attention to the role of public accountability in their conversations and relationships with audiences and funders. Public scrutiny of any institutional action had become a bureaucratic requirement of governance expected of every organization that holds the public trust, a responsibility to demonstrate, through various measures, accountability to some larger governing entity.[29]

One consequence of regulation, however, was that zoos tended to present their efforts as the conservation of wildlife or biodiversity, describing the mission of their public programs as public knowledge about the biodiversity crisis. That framing, while subject to critiques that we explore in depth in subsequent chapters, also laid the foundation for zoos confronting demands for accountability through investment in research.

In 2001, for example, the AZA initiated a three-year nationwide research project to understand *why* zoos and aquariums matter to society in the conservation arena. Funded by the National Science Foundation, the "Why Zoos and Aquariums Matter" (WZAM) project included a foundational literature review[30] and a multi-phased study of more than 5,500 visitors to twelve AZA-accredited institutions. John Fraser was an advisor on the project, the design of which included the testing and refinement of a variety of methods to measure and assess visitors' incoming motivations, interests, and knowledge, and directly measure changes in

---

[26] Hanson 2002.   [27] Morgan and Hodgkinson 1999; Davey 2006.   [28] Ranson 2003.
[29] See Bovens et al. 2014 for an overview of the history of public accountability and the relationship between government regulators and public disclosure.
[30] Dierking et al. 2002.

their learning and attitudes toward wildlife and conservation.[31] The first wave of WZAM studies (WZAM1) showed that zoo and aquarium visitors in the United States typically:

- visit more than once, though zoo research had overwhelmingly tended to focus on a single treatment for learning outcomes as instantaneous measures representing the value of a one-off intervention;
- share knowledge gain on site with people who were and were not with them at the zoo, something that had not previously been explored as a path to public literacy; and
- integrate new learnings into their lives through actions taken weeks and months after their visit; again, a topic that had not been subject to study at that time.

As such, it became apparent by the mid-2000s that changes in learning, attitudes, and behaviors could only be very partially understood through data collection that occurs at the conclusion of a zoo visit or at the exit of any exhibit. That said, visits to these institutions were shown to have positive impacts on the conservation knowledge, attitudes, and behaviors of adult visitors. Furthermore, the data show that most visitors are ready to be more engaged in conservation advocacy efforts, and zoo visits prompt them to reconsider their own role in environmental problems and recognize themselves as part of the solution.[32]

With support from the Institute for Museum and Library Services, and on behalf of the AZA, a Wildlife Conservation Society team led by John Fraser and Jessica Sickler conducted a series of complementary research studies focused on the broader social context in order to better understand what it means for an institution to be an important part of a community.[33] Their qualitative exploration of public perceptions of the role of zoos in American society began with in-depth interviews and focus groups with community representatives, visitor groups, and internal and external

---

[31] Methods included written questionnaires, interviews, tracking studies, and concept mapping activities.

[32] Fraser and Sickler 2008b.

[33] Housed at the Bronx Zoo, the Wildlife Conservation Society, formerly the New York Zoological Society, promotes the practical understanding of the science of biodiversity loss and how human behavior change can reduce negative impacts on the biosphere. Between 2006 and 2008, the organization's Public Research and Evaluation Program used social science research and evaluation to clarify various human dimensions of wildlife conservation, such as how people understand conservation concepts and what motivates them to engage in conservation activities. An example of an organization balancing field conservation programming with management of a zoo, the Wildlife Conservation Society currently manages four zoos, the New York Aquarium, and formal field conservation programs in sixty-four countries.

stakeholders. Phase two involved the development and administration of a set of quantitative surveys to establish baseline data on public perceptions across the United States. To contribute to understanding the need for zoos to be accountable for their education mission, this study identified topics that different sectors of society feel they need for information, and how the intersections of these categories might create challenges to the legitimacy of the zoo as a valued public education actor.

Museums have been described as "a transaction between a present life and the continuities of the past, a negotiation between living cultures and their futures,"[34] and the WZAM studies set the tone for a concerted and ongoing effort to actively explore why zoos and aquariums matter to people in ways that seem distinct from the perceived social value of, for example, botanical gardens or museums of art.[35] The WZAM series produced reliable data that validates the potential importance of zoos as sites for meaning-making that can change feelings and attitudes about conservation. About half of all zoogoers in the study believed zoos play an important role in species conservation and are deeply committed to animal care and education. Initiating their own questionnaire to study and begin to quantify zoos' conservation impacts in 2003, the Zoo Measures Working Group of the British and Irish Association of Zoos and Aquariums similarly concluded that assessment of the conservation impacts of zoo endeavors, including education endeavors, is both crucial and possible.[36]

In 2008, Fraser and Sickler published a national set of baseline data summarized in an accessible handbook format to support training programs and inform institutional practice. The "Why Zoos and Aquariums Matter" handbook, workbook, and visitor evaluation toolbox have been shared with zoo, aquarium, museum, and research communities as a resource to support training programs, inform institutional practice, and build the capacity of AZA member institutions to conduct their own visitor research.

---

[34] Carr 2011, viii.   [35] Fraser 2017.

[36] The 1,340 sampled adult visitors to six zoos and a nature reserve in the United Kingdom appeared to be more concerned about conservation than the public overall, though despite controlling statistically for several potentially confounding variables, the research team found that no clear and consistent measurable effect of informal education impact on adult visitors' conservation knowledge, concern, or perceived efficacy could be connected to a single visit to any of the sites studied. However, differences in the profiles of those visiting different institutions did affirm the importance of looking at within-zoo differences between arriving and departing visitors, rather than simple exit surveys across parks, to assess the impacts of different zoos. Balmford et al. 2007.

Implications

Zoos and zoo associations worldwide seek to create programming that more directly connects the public with conservation-related activities and concepts within and beyond the zoo. Yet, while offerings more consistent with how conservation mindsets and behaviors seem to develop have become more widely available, even at the time of writing we found that the default education programming and evaluation tools in most zoos continue to be based on an information transfer model of learning, even though data suggest that this approach does not speak to public audiences in the language of their values and expectations.[37]

This disconnect was an impetus for this book. A value proposition for conservation education that promotes the moral and ethical responsibilities of individuals, communities, and societies to protect and preserve wild species and wild spaces is a logical complement to existing educational programming designed to increase the knowledge of public audiences. Even very young children express analogical, imitative, and verbal identification with nonhuman species, patterns that suggest an "inner theater of moral deliberation," according to conservation psychologist Olin Eugene ("Gene") Myers, Jr.[38]

It has also been established that people who have opportunities to observe animals' conscious use of exhibit space appear to reflect on that experience using empathetic and metaphoric dimensions that can expand their worldview.[39] As institutions that have established a great deal of legitimacy regarding understanding and promoting animal well-being, accredited zoos should be known to lobby vigorously for the preservation of biodiversity – and equipping their visitors to do the same.

Through a cognitive psychology lens, the culture of nature arises through shared conceptualizations organized by human ideas and experiences. To maintain social relevance, it seems incumbent on zoo leaders to focus on "ethical and moral leadership" within public arenas wherein public audiences "discuss and debate the challenges facing society as extinction accelerates and ecosystem services are degraded."[40] In the following chapters, we will explore what available data suggest about the many ways experiences in zoo settings can influence visitors' feelings and attitudes about conservation.

---

[37] Fraser and Sickler 2008a.   [38] Myers 2007, 179.   [39] Pekarik 2004.
[40] WAZA 2015, 17.

It is a point of pride that zoos and aquariums employ so many wildlife conservation and research innovators, and are so often at the table when expertise and collective action are needed to identify or support critically endangered species.[41] We argue, though, that zoo messaging and experiences must simultaneously *help learners of all ages see themselves* as embedded within natural systems – *and see and value biodiversity as valuable, irreplaceable, and worthy of active protection and behavior change.* Zoos, luckily, are accessible venues wherein exhibit design and programming can incorporate psychology, sociology, and communication research findings, and continue testing impacts with diverse public audiences to see whether and how zoogoers appear to build new ways of thinking about species and habitat conservation.

Conservation can only occur through collective action to achieve cultural transformation as soon as possible, from wealthy and non-wealthy, and nation and individual. As currently articulated, the collective conservation mission of accredited zoos implies visitor action. What is really required, then, is focus on the pathways specifically available to zoos to transform the values, associated narratives, and action patterns that guide human relationships with the natural world. Many of the analytic lenses and findings highlighted throughout the remainder of this book, though, have potential relevance for *any* leader, educator, activist, researcher, institution, or organization seeking a deeper grasp on how people understand conservation concepts, key motivators that spark conservation action, and contexts and experiences that might accelerate the development of new social norms that will ground behavior patterns compatible with conservation goals.

---

[41] Fraser and Wharton 2007.

CHAPTER 3

Learning
Social Experiences and Captive Animals

Like every other type of museum, zoos are comfortably categorized by the general public as sites where learning occurs. Yet, zoos and zoogoers alike consistently conflate the learning function in/of zoos with zoogoers' retention of scientific facts.[1] While most zoos do offer structured on-site classes and plenty of science facts on signs posted throughout their institutions, it has become increasingly clear that this narrow conceptualization of learning inadequately captures the learning pursuits and outcomes that can and do occur in zoo settings.

Over the past several decades, learning has come to be understood as a lifelong process that is simultaneously deeply personal and socioculturally mediated; a conscious, unconscious, and continuous navigation of complex, ambiguous concepts. It has also been established that this highly personal process involves building on prior knowledge, experiences, interests, and motivations, and tends to be framed through expectations.[2] More nuanced understandings of the learning process are important because zoos hold much promise as remarkable spaces, in unique places throughout the world, where people from a broad spectrum of sociocultural backgrounds and agendas find themselves in shared spaces reflecting on abstract concepts, together.

Museum and library scholar David Carr often described library and zoo visitors as "users," noting that visitors use *their own* lenses and self-directed seeking activities to put new ideas, information, and points of reference together in ways that impose meaning and meaningfulness.[3] Indeed, in the twenty years since the turn of the millennium, a plethora of studies now show that humans in museum settings do not learn by digesting intentionally presented information, but as their personal agendas, existing understandings, and points of reference inform their encounters with

---

[1] Heimlich and Falk 2009.   [2] Brice Heath 1983.   [3] Carr 2011.

choreographed *and unchoreographed* social dynamics and cognitive and affective stimuli they experience on site. Scholars have thus concluded that learning typically occurs by design *and* by default, driven by "free-choice" experiences in informal environmental education contexts, such as zoos. We readily accept John Falk and Lynn Dierking's term free-choice, which accurately suggests that visitors choose and control their own (or their group's) goals and activities in such settings.[4] To use Carr's terminology, people who visit zoos are "users," selecting what and how they will consume presentations, information, and experiences in their own learning journeys.

Situating our understanding of the zoo experience from the perspective of a user makes it easier to recognize and explore connections between the uniqueness of captive animal display and users' priorities, understandings, and points of reference. The exercise becomes an opportunity to explore the affordances of these settings for the raw phenomena we see, and to consider how to adapt visitor experience, exhibit design, and conservation messaging for optimal conservation outcomes. Referencing a range of studies that situate the zoo as a distinct setting and tool for learning, we situate the tension between "individual-learner" narratives and the phenomenological meaning-making that occurs when individuals in groups observe live animals through the unique lens of what is important and useful in their own lives.

In this chapter, we introduce empirical evidence and personal narratives demonstrating that learning in zoos is complex, personal, layered, and social, concluding that zoogoers' motivations and emergent, on-site experiences do often converge with zoos' conservation education goals. Recognizing that increasingly nuanced understandings of learners and learning in zoo settings have already impacted, and might further impact, exhibit theory and educational programming in zoo settings, we also highlight how live animal exhibits can be approached as a conservation storytelling device that capitalizes on the nature of shared emotional experience and social learning.

---

[4] See, for example, Falk and Dierking 1992, 2000 and Falk et al. 2009. Hein 1998 was one of the first researchers to show that learning cannot be assumed to be prescribed by exhibition creators, and Heimlich and Falk 2009 note that their work builds on the thinking of previous scholars such as Maslow 1954 and Dewey 2003, who had already situated learning as a natural human process that does not necessarily require structure and intention. Dierking 2005 speaks to the dynamic and growing demand for free-choice learning, making a compelling argument that free-choice science, technology, engineering, and mathematics (STEM) learning is crucial to how people today learn and contextualize science-related knowledge and understandings throughout their lifetimes.

Layers of the Learning Process: Motivation, Meaning, and Morality

Zoos and aquariums are just one of many distinct types of tourist attraction that feature animals in captivity, and it has been recognized for decades that visitors arrive with a range of expectations and intentions.[5] Learning – an individual's highly personal process of building on prior knowledge, experiences, interests, and motivations – tends to be framed through expectations. Early phases of the Why Zoos and Aquariums Matter research program were set up to explore how zoos and aquariums serve their various public audiences. At this writing, that program is now in its twentieth year and represents a broad array of studies undertaken with a variety of research partners.

The first wave of studies, WZAM1, was led by researchers at the Institute for Learning Innovation, under the direction of Lynn Dierking, John Falk, and Joe Heimlich. With their colleagues and collaborators at numerous American zoos and aquariums, the research team specifically sought to investigate the confluence of visitor and institutional agendas to explore whether and how people situate zoos and aquariums as authorities on issues related to conservation and environmental protection. The research team's findings, published in 2007, affirmed that visitors choose to visit for multiple reasons and arrive with widely differing interests, beliefs, attitudes, and degrees of prior conservation concept knowledge. Despite these points of divergence, approximately half the sample tended to cluster around one identity motivation that directly impacted *what they did* during their visit and *what meaning they got from it.*

The grouping dubbed "Facilitators" by the WZAM1 research team described their own zoo visit as motivated by desire for a social experience that they expected to satisfy someone else, and their satisfaction was tied to this anticipated outcome. "Explorers" described their zoo visit motivation as personal interests they hoped to pursue in the zoo setting, and their satisfaction was tied to the perceived quality of the on-site learning experience, including the opportunity to see and interpret animals. Facilitators and Explorers were the most common motivational clusters identifiable amongst WZAM survey respondents; each cluster represented about 16 percent of the zoogoing sample. Somewhat common were "Experience Seekers," zoogoers who were tourists, or locals who described

---

[5] In the mid-1990s, for example, Myra Shackley 1996 noted that some visitors seek information and education, whereas others wish to be entertained.

the zoo as a valued part of their community; and self-identified "Professionals/Hobbyists," zoogoers whose visit was connected to an articulated commitment to the lifelong pursuit of further knowledge.[6]

Importantly, WZAM1 researchers found that *why* someone decides to visit a zoo not only has a direct effect on how they use the space and what they learn while they are there, but also on whether what they learn changes their ideas, attitudes, and actions. Whether a visitor hopes or intends, for example, to pursue personal fulfilment or facilitate a group experience greatly shapes that person's on-site experiences, learning pathways, and learning outcomes. To take this example further, a Facilitator may situate the zoo as a recreational environment likely to be pleasing to the children they visit with, and arrive with an explicit set of didactic aims related to describing and modeling the importance of standing in line, taking turns, playing well with others, and exhibiting concern for the well-being of animals.

The wide array of agendas now known to impact zoogoing experiences help explain why, when zoogoers were asked to describe the zoo as an institution or their own experiences in the zoo in the second wave of WZAM studies (WZAM2), few distinguished between setting and messaging – or highlighted logico-deductive science learning among the outcomes they valued about their visit. Yet, John Fraser, who served as principal investigator, notes that two key themes emerged through initial WZAM2 surveys of educators, political leaders, media, zoo/aquarium volunteers, spiritual leaders, and field biologists throughout the United States: *characterization of zoos as educational* and *characterization of the zoo as a social experience.*[7] These characterizations were valued in different ways and for different reasons, but public audiences overall, particularly parents, reported that they highly value the role of the zoo as an educational resource and setting that helps them develop empathy in their children.[8]

Whereas exhibit designers and zoo staff often conceptualize the solo visitor as an ideal type most likely to be receptive to science learning, and focus on the learning attributes of individuals,[9] zoo experiences are intrinsically social – and we now know that that most zoo visitors seek to share

---

[6] "Experience Seekers" and "Professionals/Hobbyists" were motivation clusters that represented approximately 8 percent and 10 percent, respectively, of the WZAM study sample. We note that, while 48 percent of zoogoers in the sample could be identified by a single dominant motivation, others identified with two of the five identity-related motivations described in the survey tool, and the nature and importance of overlapping motivations remains unknown. Study details, findings, and implications can be found in Falk et al. 2007, 2008.

[7] We note that this work was developed based on aggregate comparative research identified by Pekarik et al. 1999.

[8] Fraser and Sickler 2009.   [9] Pekarik et al. 1999; Falk 2006.

their experiences with friends and family. Indeed, 93 percent of zoogoers sampled in the WZAM2 nationwide audience survey reported participating in social learning experiences that involved *collaborative* meaning-making during their zoo visit, and nearly half of the caregivers sampled felt that *spending time together as a family* was the predominant intention that brought them to the zoo.[10]

Though there is significant evidence that science learning can be an important and strong outcome of zoo visits and zoo education programs,[11] visitors very rarely include "learning about science" when they describe what motivated their zoo visit. This is not to say that visitors do not care about conservation narratives or environmental issues. On the contrary, WZAM2 data show that visitors consistently report that they *want environmental information from zoos, trust zoo staff to provide that information,* and *consider the intentional development of environmental understanding to be a core responsibility of zoos.*

Respondents strongly valued the zoo for facilitating important linkages between personal connections, philosophy, and morality; they also valued the zoo for teaching visitors about animal endangerment and conservation. Unlike the public at large, three sampled stakeholder groups – zoo volunteers, teachers who had participated in zoo programs, and field biologists – perceived "learning about science" as an important zoo-visit outcome likely to lead to conservation-focused action and necessary to ground a strong conservation ethic. Respondents associated with these stakeholder groups, of course, share the common thread of repeated, work-related, personal experiences in the zoo and/or in formal science training programs. Perhaps unsurprisingly, the messaging and communication tools these respondents prioritized for zoo mission pursuit tended to reflect strong emphasis on strategies and goals for science concept and information transfer, rather than helping individuals connect to conservation concepts and practices through personal experiences of their own.[12]

While science was rarely mentioned, zoogoers in the WZAM2 samples frequently reported that their experiences in the zoo helped them establish or reinforce linkages between personal connections, philosophy, and morality that they felt were important and valuable, and many reported being intentional about using zoo experiences to develop and instill a moral foundation that includes values of care in the children they came with.[13] Emerging scholarship further affirms that *most zoo visitors want to*

[10] Fraser and Sickler 2009. [11] Fraser and Wharton 2007; Shepherdson et al. 1998.
[12] Fraser and Sickler 2009. [13] Ibid.

reflect on the moral and ethical implications of human relationships and human practices. The great majority of adult zoo visitors sampled in a follow-up study that will be published in 2020 by Heimlich and colleagues reported visiting the zoo to imagine wilderness and think about how animals and humans fit into the world – not to learn specific details about, for example, the skin color of polar bears.

Unsurprisingly, psychologists continue to confirm that animals are often significant to individuals' sense of self and morality, and that concepts of living things and perceptions of connection shape how children and adults alike perceive and interact with animals and the natural environment.[14] A robust array of domain-specific research has established that reasoning about moral obligation and moral judgement is a personal and fluid journey that begins very early in life.[15] Naturally interested in and concerned for animals, children across cultures navigate morally loaded and often conflicted cultural messaging about animals; often readily perceiving nonhuman species as feeling, thinking beings with intelligible intentions, desires, and mental states.[16] The finding that even urban children from impoverished communities have been shown to express "diverse and rich appreciation for nature, and moral responsiveness to its preservation" supports the claim that the need and propensity to affiliate with nature is a developmental universal that reflects culturally variant forms.[17] As early as 1829, Regent's Park in London was publishing guidebooks for young audiences, reflecting the long-standing attention of zoo managers on attracting children as a crucial audience segment.[18]

Conservation psychologist Gene Myers has shown that children demonstrate universal patterns of moral feelings that situate animals within their spontaneous "field of care."[19] He also highlights the importance of awe and wonder as points of entry that foster humans' emotional bonds with animals and provoke internal tensions around the moral and relational implications of human–animal connectedness.[20] So how can zoos be

---

[14] Gene Myers 2007 has used systemic empirical observation and findings from ethnographic analysis of child–animal interaction that includes cognitive, social, and language development frameworks to propose and test integrated constructs that paint a nuanced picture of children's relationships to animals. Peter H. Kahn, Jr. 1999, a developmental psychologist, offers structural-developmental theory as an alternative lens for exploring the relationship between people and their living environments.

[15] Kahn 1999.   [16] Myers 2007; Melson 2001.   [17] Kahn 1999.

[18] Baratay and Hardoiun-Fugier 2004.   [19] Myers 2007, 16.

[20] Ibid. Aldo Leopold 1887–1948, who popularized the concept of a "land ethic," similarly prioritized the sense of wonder and kinship humans can develop in response to and with regard to living creatures, as an important aspect of awareness of, respect for, and reverence toward the environment as a complex living whole that humans are only one part of (see Nash 1982, 182–199).

intentional about building from these starting points to further develop among zoogoers personal and collective understandings that connect and commit individuals, communities, and societies to the well-being of other living things?

Connections to Exhibit Theory and Design

> This observational space – the zoo – is also the constantly renewed and transformed product of the views and attitudes which it helps to shape. The staging of the zoo says just as much as the practices within it about the relationship of human beings with nature.
> – Historians Eric Baratay and Elisabeth Hardouin-Fugier[21]

The nature of zoological institutions and the exhibition philosophies used therein have constantly shifted over time in response to changing conceptions of what these sites should attempt to achieve and how best to facilitate desired outcomes. Most zoological institutions at the turn of the twentieth century had formal aims related to democratizing scientific learning about wildlife and nature through the observation of living animals. At the time, zoos typically presented straightforward displays of unusual animals that offered little taxonomic information or visual connections to native habitats. In 1907 Carl Hagenbeck introduced German audiences to a novel, naturalistic approach to zoo design that many decades later came to be widely adopted by zoos worldwide: the illusion that different groups of zoo animals, and zoo animals and zoo visitors, were not separated by physical barriers. As described in Chapter 1, cage bars were eliminated for terrestrial mammals, zoo animals were grouped by continent, and the emphasis on large, landscaped panorama enclosures gave rise to the term "zoological park."[22]

Ever adaptive, leading zoos around the world began enclosing and displaying animals in new ways, seeking to educate visitors about the look of native habitats by presenting memorable images visitors would respond to emotionally as they reflected on "the absent reality of 'the wild.'"[23] These naturalistic settings, though, rarely reflected natural habitat or species grouping contexts as they existed in nature, often featuring, for

---

[21] Baratay and Hardouin-Fugier 2004, 10.
[22] Hanson 2002. For thorough reviews of the history and evolution of the zoo enterprise, see also Kisling 2000, Baratay and Hardouin-Fugier 2002, and Hancocks 2003.
[23] Axelsson and May 2008, 43.

example, concrete representations of geologic formations and shipwreck themes.[24] Journalist Alexander Wilson and historian Nigel Rothfels have both demonstrated that the naturalistic strategy of zoo design emerged directly from a late nineteenth century romantic view of wild nature believed to have distinct nostalgic appeal.[25] As zoos shifted toward more realistic naturalistic habitat representations with a greater focus on habitat and biodiversity threats in the 1980s and 1990s, it was assumed that natural history learning would occur and a conservation ethic would develop if visitors were themselves immersed in the bucolic and multisensory landscape setting of an animal display designed to make it "intuitively self-evident" that animals have reasons for being and rights to exist.[26]

Disrupting these dominant assumptions about the construction of knowledge in zoo settings, in 1988 Barbara Birney became one of the first scholars to demonstrate that children expect to learn in zoos through unstructured social exploration. Birney found that middle school students described their learning in the zoo as curiosity fulfillment, happiness, and wonder. Few attributed what they felt they learned to conservation concepts, natural history, or their experience with exhibit design.[27] Since Birney's early challenge to the assumed nature and process of educational content gained during zoo visits, numerous museum researchers have continued to demonstrate that self-directed learning does not fit comfortably with transfer of knowledge models,[28] and a growing body of scholarship highlighted earlier in this chapter instead suggests that the construction of knowledge in zoo settings occurs as visitors situate information and experiences within the context of their own lives and existing knowledge.

Zoogoers know their learning will be sensory and include multiple modes of learning that are rarely simultaneously engaged during an intentional social outing. While vistas, smells, and sometimes unpredictable live animals provide memorable and provocative focal points for visitor engagement, such stimuli can also become barriers to zoogoers' ability to focus on the messaging intended by the institution. Sentience – the ability to sense, perceive, and interpret information from the immediate environment and

---

[24] Hanson 2002.   [25] Wilson 1992; Rothfels 2002.

[26] This claim was made by exhibit designer John Coe 1985, 2. For overview information regarding shifts in zoo design pedagogy over time, see Mullan and Marvin 1987 and Hanson 2002.

[27] Birney's research did demonstrate that children identified zoo staff as caring and good people concerned about wildlife but situated these perceptions as data points unrelated to the educational aspects of their zoo visit.

[28] Jensen 2014.

interpret those data points as emotion – is as characteristic of many animals as it is of humans, which is why sentience is often a starting point for anti-zoo critiques. Sentience also explains why human zoogoers' self-directed, on-site, in-person experiences with the animals, people, and sensory stimuli they encounter in the zoo often have greater emotional impact than their experiences looking at videos or photographs of wild or captive animals.

Centuries after philosopher David Hume suggested that emotions drive moral behavior, psychologists continue to develop deeper and more nuanced understandings of emotion as a core aspect of mental process, and conservation psychologists and zoo researchers have begun to explore the extent to which experiences that evoke "other"-oriented emotions can be leveraged to advance a conservation agenda.[29] Given that audiences respond both emotionally *and* intellectually to zoo animal exhibits – as well as to conservation information – it stands to reason that visitors are more likely to retain information and report feeling motivated to be proactive about conservation action if they connect with specific aspects of their zoo visit on a personal level. Emotion is thus a core element of current efforts to awaken and foster widespread, active commitment to the preservation of biodiversity and ecological integrity. Programming at leading zoos around the world now reflects intentional focus on offering "multiple points of contact" with animals and a framework that validates visitors' feelings of connection and compassion for animals *and* connects animals to conservation to help participants better interpret what they see on exhibit; these endeavors and the research grounding them will be a focus of Chapter 5.

Zoos have been experimenting with how to develop and share compelling stories not only about animals' needs for well-being, but also about the integration of one's identities, relational roles, and responsibilities in the many natural systems where people interact and have an impact. Beyond free-choice zoo visits, some twenty-first century audiences in diverse locales also have access to zoo experiences as wide ranging as summer and winter camps; sleepovers; early childhood programs; photography and other courses and engagement projects for families, youth, and adults; thoughtfully crafted curricula emphasizing the science of conservation; student internships; and on-site and online resources and degree programs for teachers.[30]

---

[29] Hume 1739/1978.
[30] For an engaging reflection on the history and explosive emphasis on conservation education in leading zoos in the past two generations, see Berkovits 2017.

However, it remains a challenge to overcome highly elaborated social narratives about what constitutes nature. About twenty-five years before this writing, John Fraser spent some time talking with young children about the idea of building a farm exhibit in the zoo where he was working at the time. The staff were planning the farm as a learning environment that could help introduce urban families to the idea of a human-managed landscape that also supported a wide range of wildlife. The urban children in this study were quick to name common animals that often appear in children's books – cows, goats, and horses – but domestic cats and dogs were more likely to be named first. It became apparent that the conception of various forms of landscape that support wildlife remained elusive and that farms were not considered nature by the great majority of these children, but an extension of a home environment most notably identified by terrestrial livestock and pet animals.

Spurred by findings from this unpublished zoo farm study in the Pacific Northwest, Fraser and his colleagues subsequently asked participants in a national study about family support for nature experiences in children's development. Interestingly, when asked to describe how far nature was from their home, even respondents living in rural parts of the country imagined nature as hours away from their homes.[31] This terrestrial bias suggests that zoo designs featuring simulated wild forests and deserts may be preferencing existing aesthetic tropes without necessarily pushing perceptual boundaries about the natural systems on which life depends, inadvertently reinforcing the preexisting beliefs of many zoo users that nature is a wild system found elsewhere, outside of human influence and infrequently encountered.

It is encouraging that social science researchers have now become much more nuanced assessors of how public audiences perceive, value, and learn from the human and facility assets of zoos (zoo capacities). Data related to visitors' motivations, priorities, existing knowledge, and on-site experiences can – and should – inform programming, design decisions, and the job descriptions of zoo staff, in order to deepen zoos' institutional impact as conservation actors, communicators, and social movement builders (zoo mission). Yet, better understandings of zoos' unique affordances as conservation-focused learning settings have only recently begun to gain widespread acceptance in zoo design discourse.[32]

---

[31] Fraser et al. 2010.   [32] Tofield et al. 2003.

Interpretive Tools

An early advocate for ensuring that the art and science of both exhibit design and animal care be integrated to attract visitors' attention, instill memorable impressions, and convey clear conservation messaging, architect Kenneth Polakowski analyzed the design trend of "environmental immersion" as it gained traction in the late 1980s.[33] Now prevalent and conceptually associated with learning goals, environmental immersion displays use the artifice of natural landscapes to allow zoo users to consort with animals in spaces that seem similar to the origin habitats of these species.

Well-known international zoo and aquarium architects continue to refine these techniques, drawing on material choices and personal experience with iconic habitats to create these illusions, effectively presenting the zoo as a cartoon that replicates bias and prejudice regarding what constitutes nature.[34]

We use the term "cartoon" not to diminish the value of these environments as well-thought-out display techniques, but to highlight how the scale and appearance of such displays inevitably capture only the salient dominant form of a "moment" in nature as one carefully edited viewpoint. There is, of course, nothing natural about these exhibits; they are engineered dioramas that reinforce common perceptions about animal habitats.

Exhibitions – intentionally grouped and sequenced experiences that present a thematic story – are also just one of the features visitors expect in the zoo. Zoos are distinct among museum types as settings in which visitors may find theme park and playground offerings, as well. Climbing structures, carousels, Ferris wheels, and roller coasters are neither uncommon nor unexpected. As other researchers have pointed out, sometimes the only commonality of the constructed spaces that comprise a zoo is their metaphoric wildness.[35]

Not unlike environmental immersion exhibits, signage and site maps are interpretive messaging tools created by the zoo to set visitors' structural framework and priority focus expectations. The zoo map thus becomes a

---

[33] Polakowski 1987.

[34] Jones and Jones, The Portico Group, Studio Hansen Roberts, Pat Jankowski Architect, and John Coe and colleagues, who are now operating as CLRdesign, Inc. and Ursa International, were progenitors of the environmental immersion approach and remain the most influential designers of natural landscape exhibitions.

[35] This is a topic thoroughly explored by Alexander Wilson 1992.

crucial instrument for establishing visual cues, symbolism, patterns of organization, and naming conventions that contribute to coherence while also revealing overt and subtle narratives about sociocultural norms and values. Beyond route guidance to attractions, services, and specific live animal exhibits, zoo maps juxtapose boundaries and images in ways that reflect and present socially constructed discourses about cultural heritage, relational concepts, and animals that can impact zoogoers' impressions, memories, and perceived role. Subjective and multilayered, zoo maps use various tropes that rely on cultural interpretation to impose order and organization into the zoo visitor's on-site spatial, conceptual, sensory, and emotional experiences.[36]

While environmental immersion techniques and other interpretive features were advanced as tactics to promote creative and cognitive meaning-making by zoo and aquarium visitors, and have been purported to promote learning about and greater empathy for animals and their needs, visitor data does not necessarily support these claims. As just one example of why this approach may have less of an impact than these institutions and their design teams anticipated, many aquariums use a Guggenheim-inspired spiral design that presents a linear visitor flow and sequence of engagement opportunities that feels like a film with a distinct beginning, middle, and end. Yet, a recreational fishing hobbyist who arrives with highly technical knowledge and particular interest in the visible behavior patterns of specific exotic species and how the aquarium maintains exhibit water quality may choose to bypass all on-site experiences and learning opportunities that don't seem to be directly related to those self-determined goals.[37] The different intentions, priorities, and existing perspectives and knowledge frameworks that users arrive with can and do derail or render superfluous institutional efforts to direct visitors toward proscribed learning opportunities and outcomes.

Programming Implications

Data pointing to various social role motivations that bring people to zoos and aquariums and the reasons public audiences value these institutions render moot the long-standing claim that there is no "typical" zoo visitor **and** the longs-tanding practice of "educating" and assessing zoo visitors as

---

[36] For scholarship on the topic of using cartographic analysis that situates maps as texts that reveal narratives of evolving sociocultural norms and values, see Benbow and Hallman 2008.

[37] Aquarium of Americas and Tennessee Aquarium are examples of this design strategy.

though all are alike. Indeed, the intersections of motivations, expectations, and learning in zoo settings appear to be of such importance to the social and mission value of zoos that related themes will be further unpacked throughout the remainder of this book. Clearly, though, the narrow understandings of and approaches to learning that zoos relied on for decades do not easily align with most visitors' priorities and expectations and are inadequate for the realization of zoo mission.

Findings across each wave of WZAM initiative studies have been consistent with non-WZAM studies published in the mid-oughts showing that zoogoers do learn incrementally through their on-site experiences and typically arrive with foundational knowledge of ecology and conservation principles.[38] While no evidence suggests that all zoo visitors are predisposed to, or are actively pursuing, a conservation agenda, results from second- and third-wave WZAM studies demonstrate that zoogoing audiences tend to express more deeply committed conservation attitudes and values than the general public, even if zoo users' conservation behaviors may not (yet) vary significantly from the norms in their community.[39]

Zoos and aquariums have a solid track record for public exploration of new relationships with nature,[40] and subsequent chapters situate an emerging body of research suggesting that people in zoo settings *can* experience captive animals in ways that motivate personal and collective action consistent with a conservation ethic and agenda. To this end, we argue that the unique affordances of zoos can be better deployed to realize a conservation mission when exhibit design approaches, programs, and interpretation tools are intentionally structured to support a range of motivation and learning patterns. Such considerations will better equip zoo designers and staff to facilitate a complementary array of zoo experiences that visitors who have varying degrees of preexisting conservation knowledge and different priorities may find personally meaningful.

Given that exhibits and experiences in zoo settings are often used as uniquely liminal opportunities for important conversations about care, concern, identity, responsibilities, and relational complexity, we suggest that these dialogue experiences should be understood as foundational to – rather than distinct from – conservation education goals and outcomes. Though many consider their personal morality-related agenda to be

---

[38] Wagner et al. 2006.   [39] Fraser et al. 2020.

[40] In 2007, Fraser and Wharton published an article forecasting and imagining the future of zoos, and still-relevant analysis of zoos as institutions well situated to continue a legacy of providing new conservation visions and pathways for society can be found on page 48.

distinct from the science learning process, visitors who describe moral teaching and learning motivations for their visit do often leave their visit with a stronger sense that they have decisions to make that will impact conservation outcomes. In fact, most zoogoers in the WZAM national audience survey reported feeling *they experienced a stronger connection to nature* as a result of their visit, that *their overall experience in the zoo setting reinforced their values and attitudes toward conservation,* and that *aspects of their visit caused them to reconsider their role in environmental problems and conservation action, and begin to see themselves as part of the solution.*[41] These findings confirm that zoos are well poised to credibly advance a conservation agenda because zoo visitors overall expect and are willing to learn, which is very encouraging.

Conservation *action* and idea-sharing about the future should thus be central to exhibit and program messaging, and opportunities for further learning and various forms of engagement should be clearly communicated as accessible resources and tools. We note that these concepts are distinct from the idea that a zoo might "message" a type of telling or appeal to action by a learner unaware of their options. To maximize impact, zoos should situate as activation sites and anchors within a larger conservation community and movement, rather than function as an insular network of stand-alone institutions. Relying on others as authorities is a common way of knowing by which one accepts as legitimate the information shared by someone we trust.[42] This is why adoption of new behaviors, from insisting that you not receive an unnecessary plastic straw with your beverage to carpooling, often occur as a direct result of suggestions from family or friends. As trusted authorities on matters of conservation, zoos must offer both foundational and reinforcement conservation action messaging to a wide and diverse audience. Keepers and other animal caregivers in zoo settings have particularly deep and personal knowledge of the social, emotional, and physical lives of the animals they are responsible for, and thus have unique leverage and high potential impact as conservation communicators.

Neuroscience data suggest that most learning relates to consolidating and reinforcing prior knowledge, rather than the creation of new knowledge structures,[43] so it may take time, parallel messaging or complementary experiences, and/or repeated visits to tangibly increase zoo visitors' knowledge about conservation concerns and opportunities for action, or motivation to actively pursue conservation outcomes. Given that learning

---

[41] Fraser and Sickler 2008b.   [42] Morgan et al. 1998.   [43] Dierking 2005.

occurs for different people in different ways, we note that deeper under-standings of how to advance a conservation agenda through the many potential pathways for learning in zoo settings will require multiple, creative methodologies for assessment that should sometimes include analysis of groups, rather than individuals, as the unit of analysis. Rather than situating various visitor groupings and facets of learning as categor-ically unrelated, we continue to note in subsequent chapters the overlaps that often characterize these pursuits and impact how zoogoers frame and experience learning during zoo visits, juxtaposing these findings with exhibit design theory, narrow assessments of learning in zoo settings, and zoo-related social narratives and critiques.

CHAPTER 4

Morality
Zoos As Moral Actors

One autumn morning in 2003, Sharon, one of John Fraser's colleagues at the Bronx Zoo, began reflecting aloud on why so many members of her Orthodox Jewish community visit zoos during the non-holy portion of the holidays of Passover and Sukkot. The turnstiles were, indeed, extremely busy that day, as extended families either arrived with prepaid tickets by bus from synagogues around the region or convened in the parking lot and then purchased group tickets. While these were certainly family outings, this visitation pattern was also part of these visitors' religious practice, though our zoo had no formal spaces or programming for religious gatherings.

Sharon told her colleagues that in the twelfth century, Maimonides, the Sephardic Jewish philosopher, astronomer, and physician also known as Rambam, had claimed that humans can know God through His work by observing nature, which shows the markings of Creation. Being that this was collegial dialogue, John, Sharon, and the rest of their diverse group began to discuss whether this philosophical treatise held up to the religious and spiritual heritage frameworks that may have framed their own experiences with animals and zoos. Reflecting on his studies of Daoism and Buddhism, John declared that these traditions offered little insight into the use or value of zoos because the form itself is seen as controversial for preventing animals from making self-directed choices. John also pointed out that while his scientific training could most appropriately be described as atheist, it was also undeniable that few concepts or applications could be disentangled from the very moral conundrum of whether the land management and consumption patterns that are hallmarks of the Anthropocene can possibly align with the preservation of biodiversity.

Not unexpectedly, these musings about philosophies of nature and animal observation were quickly supplanted by a joking reference to what animals show us about "the savage sexual self." Equally likely would have been a well-timed poop joke.

As inevitable as poop, innuendo, double entendre, and scatological humor can be expected at the zoo, and sometimes take legendary forms. In 2009, Jeff Swanagan, Director of the Columbus Zoo and Aquarium, very suddenly passed away in his early fifties. Jeff's death came as a shock, and many who knew him reflected together on how much he had achieved as a passionate advocate for understanding zoo visitors' perspectives.[1] While his deep commitment to zoos' conservation mission grounded all of his work, he had always approached every task with an expansive sense of humor. Jeff simultaneously shook his head and vowed to change course when he got involved in an early research project exploring public perspectives about why zoos are useful, only to discover that media messaging about and from his beloved Columbus Zoo and Aquarium had only been highlighting ... animals. There had been a such a blind spot about weaving in the context and importance of the staff who care for those animals, and the interpersonal, intergenerational, and human–animal exchanges that characterize zoo realities, that Jeff recognized an immediate need to truly, fully embrace the humorous *humanity* of zoo operations.

Jeff greenlit a clever television advertisement highlighting market research findings that the Columbus Zoo and Aquarium had recently ascended to the coveted spot of most-visited paid attraction in the area – interspersed with video of zoo staff holding shovels inside animal enclosures while reveling in fecal-numerical double entendre. This "No More Number Two!" commercial played to rapturous applause at the 2009 AZA zoo conference the year it aired. As smug smiles and pride enveloped the conference room, the closing scene cut to Jeff with his shovel and deeply sincere grin, leaving many present both heartsick and grateful to have worked alongside this inveterate joker and zoo movement insider.

Jeff's final joke, his play on the simple, the scatological, and the intersections of mortality, morality, and the indelicate, brought into focus the life lessons faith seeks to reconcile: What codes of conduct make for a good life? What are our duties of care to entities and individuals beyond ourselves? Zoos provide useful stimuli and opportunities for reflection on morality in contemporary society, and visitors value zoos for many reasons even as they recognize that these facilities receive robust moral critique related to the confinement and display of live animals.

Rumblings about zoos as chaotic incarceration have persisted since the London Zoo was first made accessible to the general public, and sustained challenges regarding the rights and inherent responsibilities of zoos as

---

[1] Swanagan 2000.

moral actors have been a persistent feature of zoo operations for the past fifty years. When John Fraser tells people that his research involves "how people think with zoos," they typically ask if zoos have fallen into public disfavor. Seldom malicious, these questions seem to arise from natural curiosity about the current status of the anti-zoo movement, often coupled with a media anecdote, rather than a personal or word-of-mouth recollection of zoo experiences.

Given that several decades of social science research on larger social narratives of zoos as symbols, destinations, and public voices with authority in civil discourse can now be drawn from, we begin this chapter with an overview of the context, understandings, and assumptions that seem to ground common zoo narratives. To highlight how metaphoric and moral critiques challenge the legitimacy of zoos as moral actors and leaders in the environmental movement, as well as some counterpoints and questions that suggest the opposite, we also take a closer look at how philosophers, public audiences, zoo and animal behavior professionals, and critics perceive, present, value, and contest the civic role of zoos in contemporary society. We introduce research that begins to upend the claim that zoos are not legitimate educational venues, an issue central to the narrative arc of this book, and note that zoo staff's knowledge of and care for animals within and beyond zoo settings can and should be central to every accredited institution's strategies to advance learning in pursuit of conservation mission goals.

Social Discourse: Evolving Narratives about the Civic Role of Zoos

Museum research has shown that visitors engage with and experience a curated museum space as a fluid and personal process of meaning-making informed by perceptions about the institution as a social actor.[2] Any in-depth analysis of zoo experiences should thus include assessment of social narratives regarding the operations of that zoo, which will often be connected to discourse about zoos more broadly.

While narratives of the zoo movement as a scientific enterprise date to the early nineteenth century, both the National Zoo and the Bronx Zoo (then called the New York Zoological Society) were established at the turn of the twentieth century with framing that clearly reflected the intention of their founding creator, William Hornaday, to ground and shape morally

---

[2] Pekarik et al. 1999; Fraser 2017.

inclusive communities that cared for animals. In 1905, a few years after it opened, the New York Zoological Society took coordinated action with the Brookfield Zoo in suburban Chicago to preserve, protect, and reintroduce a critically endangered species, the American bison, into the wild. This was an entirely new operational concept of the zoo as a balancing force seeking to counter the human destruction of wildness, and it changed internal and public narratives about the social role of zoos in a time when it was becoming increasingly clear that places untroubled by human impact were no longer abundant. Such narratives ultimately supported the shifts of perspective that led to the creation of national parks as protected landscapes that could be protected as artifacts less sullied by human occupation and manipulation.[3]

Zoos have never, however, had an exclusively positive relationship with the environmental movement.[4] Despite the noble intentions of their original founders, zoos in the 1900s struggled for decades with prurient and hedonic usership and management practices. Even as the more strident of their leaders laid claim to scientific study, this period also featured anthropomorphic presentations of chimpanzee "tea-parties" and elephant shows that invoked a circus history. Furthermore, until the 1970s these institutions relied heavily on the ongoing supply of animals from the wild and had a tragic record of sustaining the animals and species in their care. The disconnect between appeals framed around the childlike innocence in animals and young visitors, purity narratives around noble intentions to support learning, and the zoos as consumptive users of animals did not go without notice in public discourse.[5]

Zoo-As-Chaos Metaphor

Among the rhetoric and social narratives familiar to public audiences in the West are metaphoric connections that challenge the legitimacy of the zoo as a valuable force for habitat conservation or conservation education. Zoo, as a metaphor, is in fact commonly used to connote disorder and confusion, a usage in distinct contrast to the ordered assembly of animals and man reflected in the Judeo-Christian narrative of Noah's ark. Historical analysis suggests that while the menageries of the past *were* typically viewed as "disorderly," the blending of scientific research, education, and entertainment endeavors to create "authentic" experiences of nature in modern

---

[3] Fraser 2004.   [4] Morris 1969/1994; Kellert 1996.   [5] Diamond 2014.

zoological parks also creates various tensions that require negotiation by zoo managers and zoo visitors alike.[6]

Perhaps it is a nod to both traditions that the term "zoo" is used pejoratively within organizations to explain emergent or ongoing situations perceived to be illogical, chaotic, and uncontrolled. In the early 1990s, Julie and Kenneth Kendall sampled sixteen businesses in the United States and found that this common, frustration-invoked metaphor is used to imply a system facing no apparent external danger that experiences tumult and those affected perceive no constructive top-down direction or solution.[7] Implicit in this metaphor, then, is the presence of a (perhaps arrogant) controlling zookeeper loosely fulfilling maintenance responsibilities, rather behaving as a visionary leader using appropriate organizing principles. A decade after the Kendalls published their findings, Judith Lloyd Yero similarly found that the zoo-as-chaos metaphor was in ubiquitous use among teachers in the United States, concluding that statements such as "my classroom is a zoo today" are widely understood in Western culture to suggest the tense dynamic of students behaving in a negative, wild, and uncontrolled manner – and lacking free agency.[8]

Cultural critic John Berger claimed in 1980 that a long cultural legacy of blurring animals' distinctions to make them reflections of the world people wish to inhabit has made zoos in the United States little more than a disappointing mnemonic reinforcing human dominion over nature by reducing animals to spectacle in contexts where people have driven most species out of the physical environments they inhabit – and many to extinction. Berger thus claimed zoos have a narrow purpose as a memory device for recollecting how we wish to believe the human-created world is organized, now that few people today have regular experiences with wild animals living freely in nature.[9] We note that characterization of zoo animals as objectified "others" and the widespread use of zoo metaphor to imply chaotic arbitrariness and the absence of agency are connected underpinnings of anti-zoo arguments that may impact the ability of zoos to claim legitimacy as leaders in biodiversity conservation.

---

[6] Hanson 2002, 9.   [7] Kendall and Kendall 1993.   [8] Yero 2002.

[9] Cornelius Holtorf 2013, another scholar who has explored the zoo as a metaphorical place and realm of nostalgic memories, similarly notes that zoos, in conjunction with zoogoers, evoke the past within a very specific and consumable culture of history. Given this reality, Holtorf claims that decisions about zoo design and messaging can and should involve grassroots negotiation, as local perspectives add fascinating nuance and personal relevance to zoo experiences. We agree.

Table 4.1 *Distinct topic areas that impact public trust in zoos and aquariums*

Categories	Topics
Ethical integrity	Ethics
	Inform about specific animals
Conservation agency	Wildlife agent, informant, activator
	Collaborator in conservation
Transparency	Advise on sustainability practices
Quality	Quality attraction
	Quality experience

Source: Knology 2019

The Ethics of Wild Animal Display

Ethical precepts and ideas about rights are human concepts that are never tidily defensible. From antiquity, captive wildlife displays have demonstrated how governing cultures value animals and represent their attitudes toward nature.[10] The popularity of zoos offers little evidence to combat the moral arguments raised against these institutions.

Although zoos and aquariums are perceived to play a pivotal role in wildlife conservation, including through the education of public audiences, media depictions of zoos and aquariums that emphasize animal captivity may erode public trust.[11] Knology, the think tank with which both authors are affiliated, recently undertook a systematic survey of organizational trust in zoos and aquariums, contrasting how people perceive the current performance of these institutions against their expectations for establishing trust. Study data point to seven distinct ways the public assesses their trust in zoos and aquariums (see Table 4.1).[12] Zoos were reported to be highly trusted, though a trust gap did remain. We found that the "zoos and aquariums" category was unique when compared to other types of organizations, with care and stewardship of animals an issue that shifted the balance on expectations. The largest disparities between perceptions and expectations were for items that assessed ethical integrity – how well zoos and aquariums maintain and communicate animal welfare. Other research has shown that the public overall erroneously assumes that

---

[10] Mullan and Marvin 1987, Rothfels 2002, and Grazian 2015 are among the many authors who have written extensively on this issue.
[11] Fraser and Sickler 2008b.   [12] Rank et al. 2018.

government sets higher standards for animal care than zoos with self-policing associations, and base perceptions about animal welfare on physical and mental standards for different species that do not necessarily rely on space, though abstractions about space are often heavily weighted.[13] The ethical burden of transparency in the management of wildlife seems to overwhelm any conceptual distinctions between quality assurance and legal compliance that are typically part of how people think about the degree to which they trust an airline, a manufacturer, or a hospital. Seemingly in this vein, the public trust gap highlighted in this study had little to do with actual institutional performance, but did appear to reflect common misconceptions that animals in nature roam without constraints; zoos cannot ever appear large enough to replicate such imagined conditions.[14]

An oft-cited edited volume published in 1995, *Ethics on the Ark*, offered robust critical analysis of the various functions of the zoo and a wide range of opinions about ethical implications related to how these institutions manage their living collections, goals, and operations. Reinvigorating Fairfield Osborn's claims and intentions thirty-plus years later, a chapter by Terry Maple, Rita McManamon, and Elizabeth Stevens cites a zoo's ability to motivate visitors to think about *and* act for conservation as a criterion that differentiates a "good zoo." Several of the contributing authors collectively recommended that best practices supported by Osborn become the industry norm: strategies to change attitudes and lifestyles; inculcate respect and concern for wildness; and develop exhibits that concentrate on concepts and ecosystems, not just species. The editors noted widespread agreement that efforts to protect species and ecosystems are justified, but significant disagreement about which methods are justifiable and appropriate.[15]

We seek to identify the logic grounding some of the more common negative portrayals of zoos and what evidence would be required to advance reasonable dialogue regarding the role of live animal displays in contemporary society. It is beyond the scope of this book to explore the ethics and tensions around bioengineering, the culling of "surplus" captive animals, and other highly controversial practices that occur in some zoos; we focus only on the four basic moral argument structures that tend to characterize broad-stroke anti-zoo narratives today:

1   Claims that zoos are immoral for denying freedom to sentient beings (*utilitarian or rights-based philosophical critiques*).

---

[13] Clark 2013; Shepherdson et al. 2013.   [14] Rank et al. 2018.   [15] Norton et al. 1995.

2 Claims that zoos inhumanely exploit sentient beings for commercial profit, disregarding captive animals' emotional, physiological, and/or psychological needs (*science-based critiques of captive animal care and wellbeing*).

3 Claims that zoo displays misrepresent animals and nature, and thus invalidate the concepts of wildness, wilderness, and education that are ostensibly at the heart of the zoo enterprise (*authenticity critiques*).

4 Claims that zoos diminish children's moral reasoning about care for nature (*moral development critiques*).

Utilitarian or Rights-Based Philosophical Critiques

Prominent utilitarian philosophers such as Jeremy Bentham, John Stuart Mill, and Peter Singer have claimed that sentience, the capacity for suffering also experienced by humans, entitles nonhuman animals to the same moral considerations and obligations as humans, not as a right, but as the logical and ethical pathway to maximizing the overall happiness of beings able to experience happiness.[16] Like other philosophers who offer rights-based justification for animal liberation claims, Tom Regan, points out that if sentient animals have intrinsic value – an absolute moral right to respectful treatment because they share morally relevant features with humans – the utility of sentient captive animals for human ends becomes completely irrelevant.[17]

The position that animals should have moral importance in their own right if they are provably sentient beings is, similarly, the starting point of scholars such as Dale Jamieson and Randy Malamud, who recognize a wider range of animals as sentient and propose that zoos are categorically immoral for denying sentient captive animals' inherent right to self-

---

[16] In the late eighteenth century, Jeremy Bentham presented a thoughtfully reasoned argument that utilitarian principles – a moral framework in which actions are judged based on the degree to which they increase pleasure or decrease pain – should be applied to animals. John Stuart Mill, subsequently defending utilitarianism as a guide to practice, acknowledged animals' limited capacity to reflect on and anticipate impacts of their actions on other beings' happiness but pointed out that animals can *experience* happiness and are thus worthy of consideration. More recently, Peter Singer 1975 used a utilitarian approach to determine that the always controversial line between living things that are and are not entitled to moral consideration can arguably be drawn at the level of advanced and less advanced invertebrate species, "somewhere between a shrimp and an oyster."

[17] Deontologist Tom Regan 1983 argues that the only animals that qualify for moral consideration express beliefs, desires, preferences, emotions, self-consciousness, memory, a sense of the future, and an emotional life; given these self-established criteria, the only animals Regan personally believes have intrinsic value are mammals at least one year of age.

autonomy.[18] Yet this line of deductive argument is weakened by the fact that *non*-captive animals have no inherent rights, natural habitats being replete with predators, parasites, competition, harsh weather events, and other such obstacles to survival, freedom from suffering, and free movement.[19]

Singer, interestingly, does not base his insistence that captive animals are entitled to liberation on a claim that animals (or humans) have (or should have) intrinsic rights, but on the idea that any form of discrimination against any sentient being is ethically wrong. Singer's is also an extreme position, though: If we accept no moral distinctions between species, we cannot prioritize our desire to be free from mosquito bites over the mosquito's right to do what mosquitos do.

John Fraser notes, for example, that his late uncle Vern, a salt-of-the-earth farmer and steel worker who disdained philosophical arguments that seemed distant to the working class, was a right-wing conservative who held strong ethical positions on many topics. Despite being an animal lover devoted to his prize-winning show horses and the beef cattle on his farm, Vern would have had no tolerance for any of these critiques. While the cattle he raised were cared for and known for their unique personalities and roles they played in their herds, Vern identified as a utilitarian who limited the rights of animals to what he felt were their perceptual limits. Vern's death coincided with Mary Temple Grandin's first revolutionary assessment of how to redefine abattoirs, though we imagine he would have appreciated this scientist's practical view of animal sentience and animals' rights to avoid suffering; Vern certainly shared Temple Grandin's views regarding the joys animals experience being and consorting with others.[20]

Science-Based Critiques of Captive Animal Care and Well-Being

The second anti-zoo narrative thread also reflects claims about sentient animals' rights to free agency, but on the basis that zoos do not adequately provide for captive animals' psychological and physiological needs. Indeed, many critics articulate the need for zoos to reconcile their care practices beyond anthropocentric values.[21] While some physiological or psychological needs critiques contest the legitimacy and appropriateness of all zoos,

---

[18] Malamud 1998; Jamieson 2002.   [19] Wagner and Kenski 1988.

[20] Grandin 1980. Grandin's animal behavior studies identified stressors and threats related to cattle herd behaviors. Her studies eventually led to the creation of more humane processes for animal slaughter.

[21] See, for example, Mather 2001.

categorically, others distinguish between the ethics of zoos that exercise moral concern for animals and zoos that do not.[22]

Some science-based critiques claim that sentient animals have an intrinsic right to feel well through freedom from prolonged or intense fear, pain, and other negative states; function well in terms of physical health and normal growth and functioning of physiological and behavioral systems; and also lead "natural lives" through the development of their natural capabilities and adaptive capacities.[23] Critics who categorically oppose zoos sometimes liken zoo captivity to animal prison.[24] Some animal behaviorists, for example, argue that the artificial structure of zoo animal social groupings and space constraints cannot meet animals' needs and are thus unethical.[25]

Critics and visitors who distinguish between the morality of good and bad zoos, on the other hand, typically seek to measure the degree of care a zoo accords its captive animals and what conditions are necessary to ensure that care needs are met.[26] In general, such critiques are defined within a regulatory framework that addresses the intentionality of animal caregivers or those who would regulate their activities within a given culture.[27] Empirical evidence amassed by animal behavior researchers tends to be the basis for judgements about animal welfare and about animal management and care decisions in zoo settings.[28]

Animal welfare has been a point of tension and contestation by those who manage and work for accredited zoos, too.[29] The physical and behavioral health of captive animals were not understood as interrelated until the 1980s.[30] While Hal Markowitz called on zoos to integrate

---

[22] Norton et al. 1995.

[23] Adaptative capacities refer to an animal's ability to maintain itself independently and in a state of reasonable well-being in an environment suitable for its species, acknowledging that different species have distinct needs and adaptive intelligences, and different abilities to modify behavior for a wide range of situations including, potentially, captive environments that provide sufficient opportunities for the animal to employ natural capacities (Musschenga 2002).

[24] Numerous scholars have explored symbolic and metaphoric claims that zoos are institutions that, like asylums or prisons, reflect and reinforce power relationships through designed environments that overexpose and punish, ideas connected to the theoretical work of Michel Foucault and Jeremy Bentham. See, for example, Morris 1969/1994, Berger 1980, and Mullan and Marvin 1987. Concluding that zoos do not practice power through panoptic design, and that punishment rhetoric does not often match the experiences of captive animals, Braverman 2013, 86–91 notes that panoptic and exhibitionary affordances in zoos actually appear to have overlapping institutional purposes: conditioning zoo visitors to connect with and internalize an ethic of care for the displayed animals, and zoo staff to see those same animals as beneficiaries of scientific governance made possible by the institution of captivity.

[25] Bekoff 2003.   [26] Woods 2002.   [27] Hancocks 2003.   [28] Shepherdson et al. 1998.

[29] See Fraser and Wharton 2007 regarding morality concerns raised by zoo managers.

[30] Burghardt 1996.

behavioral enrichment and training into animal display design and operational norms in 1981, this active intervention approach was slow to take
hold until animal behavior research began to emerge as a field of study.
Opportunity-rich animal environments that made enlightened animal care
and caretakers, as well as animals' innate competencies, more visible to the
public became more mainstream after zoo visitors and staff alike began to
voice concerns about the quality of animal presentation and animals'
quality of life within traditional enclosures.[31]

Second Nature, a 1998 follow-up to *Ethics on the Ark*, sought to identify
"common ground that unifies zoo and aquarium biology and animal
welfare."[32] Contributing authors from the fields of animal behavior, zoo
biology, and psychology shared data and emerging best practices related to
the use of food-based, physical, social, cognitive, or sensory stimulation
strategies designed to counter some of the negative aspects of confinement
through a more humane management culture committed to the psychological well-being of captive animals. Operational standards that included
such "environmental enrichment" approaches (also called "behavioral
enrichment"), they collectively claimed, would more appropriately reflect
new understandings of the complex physical, social, and psychological
dimensions of animal lives – and simultaneously provide more informative
and interesting exhibits to advance a conservation education agenda.[33]

Animal welfare science quickly became a focus and commitment of
accredited zoos in the United States, and accredited zoos must now ensure
their captive animals are stimulated through training and other stimuli
designed to improve physiological and psychological well-being and prevent the individual animal from visible displays of "unpleasant states such
as pain, fear, or distress."[34] Commissioned in 2000, the AZA's Animal
Welfare Committee was charged from the start with assuring the public
that zoos are good and ethically legitimate places for animals, in addition to
supporting member zoos in the identification and application of best
practices in animal welfare.[35]

Much like the zoo-as-purposeless-chaos metaphor, critiques of captive
animal care and well-being assume that animals' lack of free agency in zoos

---

[31] Coe 1997.   [32] Shepherdson et al. 1998, xiv.

[33] A strong incentive to support research and strategy related to enrichment for captive animals was the
emergence of data about stereotypic behavior, repetitive patterns with no apparent goal that can be a
coping mechanism or outward display of distress (possibilities that point to suffering related to
mood and/or environmental conditions), or a relatively harmless habit, according to Musschenga
2002.

[34] www.aza.org/animal_welfare_committee.   [35] Hanson 2002.

is not comparably balanced by the care and comfort experienced by the animals held captive, social and cognitive gains for zoo visitors, and motivation and knowledge to act in ways that advance a conservation agenda. Welsh philosopher Mark Rowlands, as one example, establishes that a wide range of the animals displayed in zoo settings produce natural opioids to suppress pain and stress. His conclusion that zoos are unjustifiable is based on two assumptions: forced confinement causes sentient animals to suffer; and "zoos are neither necessary for, nor effective at," the promotion of education or scientific research of vital interest to humans.[36]

The extent to which zoos can legitimately claim their displays of captive animals impact visitor learning – and, by extension, conservation outcomes vital to the survival of animals and well-being of humans – are, of course, core topics of this book, and in the final section of this chapter we introduce empirical data that show important learning does occur in zoos. With respect to scientific research in zoos, while analysis of the moral implications of such endeavors is beyond the scope of this book, we do note that scientific research in zoo settings has significantly impacted human understandings of animals' complex cognitive capacities and mental states. It has also been shown that deeper understandings of animals' complex cognitive capacities and mental states have impacted social narratives around perceived moral responsibilities to wild and captive animals – and impact the degree of compassion public audiences express toward different species.[37]

For example, despite the fact that living collections in aquariums (but not zoos) are frequently stocked and restocked from wild populations, aquariums are perceived as less controversial settings for keeping and displaying live animals because the public seems to believe fish are "inherently less aware of their restricted surroundings."[38] Indeed, the "complex milieu of logical and emotional factors" that tends to motivate captive animal care and well-being critics[39] may also contribute to the broad appeal of written and visual work about animals that integrates emotion, reason, and morality. Diana Reiss' *Dolphin in the Mirror* (2012), Lawrence Anthony's *The Elephant Whisperer* (2009), and Jennifer Ackerman's *The Genius of Birds* (2016) are examples of widely read novels that weave

---

[36] Rowlands 2002, 159. Assumptions about suffering in the animal rights movement are often based on human notions of suffering (anthropomorphism), as well as a preference for cultural relativism, the idea that each culture should treat its animals and people equivalently, but not be held to the rights standards of other cultures because the context is different; see Sutherland and Nash 1994.
[37] Herzog and Galvin 1997.   [38] Shackley 1996, 111.   [39] Herzog and Galvin 1997, 237.

personal experience narratives into scientific descriptions of the emotional and intellectual capacities of specific types of animal to provoke concern and reinforce the need for widespread conservation action.

Because the minds, relationships, and emotions of animals fascinate people and help us understand and connect to animals, we feel zoo staff's knowledge about and care for animals within and beyond zoo settings should be embraced and leveraged as a dimension of learning. Zoo staff, keepers especially, have a unique opportunity to spark dialogue about human–animal relationships and conservation ethics because zoogoers value keepers' knowledge about the animals on site and how best to care for them. When she had a chance to spend time in zoos (and with zookeepers) observing, reading about, and reflecting on how captive animals are cared for and trained, journalist Amy Sutherland discovered so much about animal and human behaviors that she found she had enough content to write what became a popular novel, *What Shamu Taught Me about Life and Love and Marriage* (2008). This key finding is one to which we will return in other chapters.

Authenticity Critiques

Several scholars have argued that the power dynamics of captive animal display disqualify zoos as sites of valid education about animals and nature. Authenticity critiques challenge the institutional legitimacy of zoos as sites that perpetuate inappropriate/inaccurate representations of nature in ways that are counterproductive to the goal of providing visitors with learning experiences that advance conservation.

Philosophy professor Ralph Acampora proposes that the over-exposure of endangered animals in zoo displays makes them less rare to the average person and that this inauthentic representation is akin to pornography and results in a dominance narrative inconsistent with valuing animals and wildness.[40] Ecocritic and cultural analyst Randy Malamud criticizes zoos as decontextualized conveniences that reflect a degraded cultural imagination, concluding that animal confinement and display cannot be compatible with imparting an ethically defensible conceptualization of humans' relationship with and responsibilities to animals.[41] Alan Beardsworth and Alan Bryman have similarly proposed that the commercialization of zoos through merchandising and theming has made zoo settings hard to distinguish from inauthentic or fictional models of nature.[42] While

---

[40] Acampora 2005.   [41] Malamud 1998.   [42] Beardsworth and Bryman 2001.

Beardsworth and Bryman are concerned that objectifying and displaying animals in zoos can decrease visitors' sympathy for animals, social ecologist Stephen Kellert has claimed that zoos promote *negative* relationships with animals when live animal displays are frivolous and force separation.[43]

In 2005, Keekok Lee proposed that the ontology of zoo animals creates a first degree of separation because captive breeding and development occur outside the evolutionary pressures of natural environments. Challenging assumptions about meaning-making, learning, and values development in zoos, Lee describes how the immured conditions to which captive zoo animals are subject create a human-defined "wildness" with no resemblance to wild conditions. She claims that live displays cannot be considered or presented as a simulacrum of wildness or authentic environmental conditions, and categorically opposes zoos as sites that cannot legitimately educate visitors about nature.

More recently, Irus Braverman published two books that similarly situate zoo animals as domesticated or bioengineered versions of wild animals, rather than the "wild life" zoos purport to house, a lens that validates concerns about the counterproductive objectification of wild nature in zoo settings.[44] Like Lee, Braverman notes that zoo-born offspring have a different ontological status than animals in the wild, become evolutionarily distinct, and cannot contribute to the wild genetic pool of their species; she thus claims zoo-born offspring cannot be legitimately promoted as tokens of a wild species. While Braverman situates zoos as institutions practicing control through care, she believes zoos have an important civic role of caring for animals and educating the public to care about animals and nature. Lee, on the other hand, argues that zoos are not legitimate actors advancing a conservation agenda through captive breeding or education that provokes conservation behavior, and that these institutions must thus refocus their mission on only the important cultural task of providing families with "good wholesome recreation and fun."[45]

The authenticity criticisms outlined here question the legitimacy of the zoo as a cultural institution on the basis that live animal displays cannot deliver the interpretive stories and conservation value they claim to provide. Clearly, the civilization–wild nature dualism zoos leverage to draw crowds by marketing "wildness" is also a powerful tool for anti-zoo critics who claim that "wildness" implies moral responsibilities, including, perhaps, the obligation to allow the captive animals featured in live displays the right to *be* free and wild. However, and in rebuttal to concerns that

---

[43] Kellert 1996.   [44] Braverman 2013, 2015.   [45] Lee 2005, 117.

inauthenticity may reduce visitors' attention to and concern for environmental issues, it has been established that zoo members typically self-report
more concern about the environment than nonmembers.[46]

A few years ago, adult visitors were surveyed at AZA institutions to
assess their mindset regarding whether animals should be kept in zoological institutions. The findings speak to the extent to which zoos' institutional credibility is connected to communicating how zoos pursue and
achieve conservation mission impacts. Whether each respondent had, or
had not, received information about the conservation work being carried
out by member institutions in the AZA network had a dramatic impact on
response patterns. Among visitors in the study sample who received no
such description, 25 percent reported no objections to keeping animals in
zoos or aquariums, 51 percent reported some objections, and 24 percent
categorically objected. Among visitors who did receive the description,
69 percent reported no objections, 26 percent some objections, and
5 percent categorically objected.[47]

Unfortunately, the characterization of zoos as tourist attractions and
research venues has been shown to further complicate the complex undertaking of public interpretation,[48] and we certainly agree with Keekok Lee
that zoos should stop marketing oxymorons such as "wild animals in
captivity." We note, however, that authenticity concerns tend to lack
merit when "education" and "education goals" are understood, pursued,
and measured in more robust ways. Rather than approaching and describing learning as a narrow process of information transfer, approaching
visitor learning as a broad range of self-directed personal learning experiences allow zoos to research a much wider assortment of outcomes that can
be leveraged to refute key arguments of zoo detractors.

Through such a lens, in fact, we find it completely acceptable for
enrichment tools to stand out as very distinctly human objects within
naturalistic live animal displays because we know that most visitors question unexpected or dissonant elements, and questions often lead to learning. John Fraser has seen how enrichment features as seemingly random
and clearly artificial as a large plastic pickle, installed with purpose in a live
animal exhibit, catch people's attention, certainly, but also elicit responses
and provoke conversations that reflect layered narratives of learning about
animals' psychological needs and human relationships within and toward
the biosphere. Importantly, such presentations go beyond a natural history
focus and do not perpetuate the perception that all traces of human

---

[46] Clayton et al. 2011; Clayton et al. 2014.   [47] WAZA 2015.   [48] Fraser and Wharton 2007.

influence must be absent (or at least invisible) for wildlife to enjoy positive physical and cognitive well-being.

Moral Development Critiques

Peter H. Kahn, Jr. leads the Human Interaction with Nature and Technological Systems Lab at the University of Washington. He has written extensively on children's moral development and how children's processes of moral development are being impacted by emerging technologies. Kahn was also an early collaborator in the founding of conservation psychology and has studied the relationship between child development and experiences in nature. Troubled by children's experiences at zoos, Kahn has taken the position that zoo visits are counterproductive for a child's moral development.[49] His argument is based on the study of children's independent reasoning around the duty of care owed to animals and how children reason through animal needs. Kahn's psychology work appears to parallel the philosophical arguments made by Ralph Acampora, but points to different causal factors. Kahn's suggestion is that children are likely to interpret zoos as authority figures that permission the oppression of animals for human pleasure, inferring anthropomorphic needs in ways that diminish children's moral reasoning about care for nature and potentially lead children to believe that artificial displays accurately represent nature.[50]

While Kahn's research does elicit this feedback when children are extracted from their social group and tested, we challenge this line of inquiry in light of research from the past two decades revealing that children do not visit or make meaning from zoos and aquariums in isolation. Children typically experience the zoo under the care of a known adult or teacher whose interventions tend to focus specifically on propriety and questions of moral responsibility. We do not discount the importance of children's independent reasoning in these settings and agree that zoos create an imposed order and moral questions worthy of deliberation, but note that zoo research findings highlighted throughout this book show that while on-site exhibitions are liminal devices open to interpretation and reasoning by all users, guidance from caregivers and zoo staff greatly informs children's navigation of moral responsibility and locus of control considerations during zoo visit experiences.

---

[49] Kahn 2009.   [50] Kahn et al. 2008.

Education Benefits: Paths to Deeper Legitimacy

Most people in Europe and the United States appear to have found it self-evident for decades that educational efforts are crucial and justifiable pursuits in any zoo.[51] Even so, the vast diversity of the facilities that fall under the broad "zoo" umbrella and widespread use and acceptance of the zoo-as-chaos metaphor mean there continue to be sites *and* social narratives from which critics mount plausible arguments against zoos as legitimate social institutions. It is only recently, though, that any argument *for or against* zoos could be supported by robust empirical data that has ethical (as well as operational and pedagogical) implications.[52]

WZAM surveys in the mid-2000s, for example, confirmed that, before visiting a zoo, members of the United States public tend to personally reconcile competing narratives about (1) what zoos and aquariums can and should be and do; (2) how these institutions represent dominant social views; and (3) the relationship of these institutions to environmental values, compassion, and public ethics. In contrast with common critiques of zoos' animal care, education, and conservation legitimacy, a vast majority of respondents identified zoos and aquariums with the dual mission of education and worldwide conservation and felt that zoos and aquariums adequately care and provide for the animals in their care. Only 9 percent of respondents reported feeling adamantly that zoos are inhumane and animal captivity is wrong; 6 percent felt this way about aquariums. Another 4–7 percent of respondents felt sympathetic to these ideas. The data further suggest that approximately half the public strongly believes zoos and aquariums seek animals' holistic well-being; more than half seem to regret that zoo animals are not wild but feel it is not immoral to keep and display animals in these institutions; and approximately half feel captive animals can still be considered "wild." Very few respondents agreed strongly that zoos and aquariums have no educational value, and the data highlights how consistently the news media "authorizes" zoos to provide the public with educational information about the lives and care of animals and human–wildlife interactions.[53]

---

[51] Baratay and Hardouin-Fugier 2004; Fraser and Sickler 2009.

[52] Noting that additional research was needed to clarify what social relevance accredited zoos in the twenty-first century can legitimately claim, scholars in the early aughts began to propose agendas for zoo research that might directly address anti-zoo critiques by testing, for example, whether living animal displays in accredited zoos offer unique attraction attributes, whether zoos support public enlightenment, and how humans' zoo experiences impact zoo animals. See, for example, Mason 2000 and Davey 2006.

[53] Fraser and Sickler 2008b.

To lay the groundwork for more nuanced understandings of whether, and under what conditions, zoogoing influences measurable mission-related benefits, sociologist Eric Jensen developed a mixed methods study of children aged 7–15 years before and after an educator-guided or unguided visit to the London Zoo. When the data showed that interpretive signage as a standalone approach to visitor education is inadequate for achieving measurable learning outcomes for the focal age cohort, this finding was quickly embraced by the anti-zoo movement. Yet, Jensen's data suggest that zoos show much potential educational value for young visitors. Rather than showing that learning does not occur in zoos, Jensen's findings seem instead to point to the crucial role of conservation educators as "toolmakers" who develop conceptual resources that tend to deepen children's understandings of science.[54]

With colleagues Markus Gusset and Andrew Moss, Jensen subsequently initiated the first large-scale international impact evaluation study exploring whether zoogoing directly impacts biodiversity and conservation literacy. Aggregate results of these scholars' pre- and post-visit surveys of 5,661 visitors to twenty-six zoos and aquariums in nineteen countries offer compelling evidence that experiences in these institutions significantly increase visitors' understandings of biodiversity and identify individual-level actions they can take to help protect biodiversity.[55]

Building on this finding, Jensen, Moss, and Gusset initiated a follow-up study to establish quantitative evidence pertaining to the long-term educational impacts of zoo and aquarium visits. Two or more years after they had visited a zoo or aquarium and completed pre- and post-visit surveys, respondents' (comparatively higher) post-visit understandings of biodiversity were shown to have remained the same, and respondents' knowledge of actions to help protect biodiversity had continued to deepen. Though small-scale, this longitudinal study tracking individual-level learning impacts over an extended period suggests that the immediate positive biodiversity and conservation learning effects of zoogoing appear to be long lasting, and that zoo experiences may lay the groundwork for ongoing learning.[56]

While measurable learning outcomes are but one aspect of the moral complexity of zoos and zoo operations, these findings are a blow to the authenticity critique that zoos are not educational venues. Data showing long-term biodiversity and conservation learning impacts, including increased knowledge of practical pro-conservation actions, further solidify

---

[54] Jensen 2014.　　[55] Moss et al. 2015.　　[56] Jensen et al. 2017.

zoo industry claims of social relevance and civic value. Jensen's finding that zoo staff and volunteers who assume the role of conservation educators help children develop conceptual resources they can subsequently build on within and beyond the zoo setting draws further attention to visitor engagement with staff as a dimension of learning that can and should be embraced to advance conservation mission goals. Subsequent chapters reveal additional research findings that point to learning that occurs in zoos, as well, touching on another thread of anti-zoo critique: Can zoos legitimately claim that specific conditions and processes unfold on site that appear to support the translation of conservation learning and practical knowledge into conservation action?

CHAPTER 5

Pleasure
The Educational Leisure Value Proposition

While natural history museums, art galleries, and zoological gardens all feature collections curated for presentation and public interpretation, the former settings are typically considered high culture destinations, whereas zoo and aquariums have no such cultural status and are often situated as leisure venues for "mere recreation and amusement."[1] Even zoo staff and politicians, furthermore, trivialize fun and pleasure as obstacles to the very serious endeavor of conservation education.[2] Should we be so quick to assume that fun is the opposite of learning? The two are orthogonal concepts, not opposing end points on a linear scale. Learning can be fun – or, when it isn't, is more like, to borrow from blogger and teacher Dan Kent, "like drinking old milk from a dungeon bucket."[3] To situate where the fun to be had at the zoo sits within mental process, this chapter explores the ideas of pleasure, enjoyment, fun, leisure, and recreation from their conceptual definitions and through an amazing amount of data that seem to render erroneous the assumption that the pleasure sought and derived from a visit to the zoo is somehow inconsistent with the pious conservation mission these institutions espouse.

Since the turn of the century, psychologists, sociologists, and anthropologists have dedicated a serious amount of time to the study of fun. Scholars have pulled fun apart in many ways and even learned that fun is a life-giving force *of* nature. We begin with an overview of how sociologists and psychologists frame pleasure-seeking leisure behaviors, and subsequently introduce recent research that clarifies several factors that appear to impact how adult visitors evaluate personal enjoyment during a zoo visit. Emergent understandings of which factors in zoo settings tend to contribute to zoogoers' enjoyment then allow us to reimagine how these

---

[1] Mullan and Marvin 1987, 117. [2] Fraser and Sickler 2009.
[3] https://medium.com/@thatdankent/learning-is-not-fun-64ed23f63d9d.

same factors can also facilitate visitors' awareness of and perceived connections to a conservation agenda.

For starters, sociologists distinguish *pleasure/enjoyment* as a personal positive feeling and *fun* as collective satisfaction, though these concepts are often used synonymously and are frequently intertwined.[4] English professor John Beckman provides a thought-provoking critical analysis of fun in his book *American Fun: Four Centuries of Joyous Revolt*, describing fun as a fundamentally elusive but extremely important concept in human understanding. Unlike pleasure/enjoyment, which can be felt, entertainment is passively consumed and predominantly vicarious. To put this more succinctly, fun is active enjoyment that must be *had*.[5]

Hardly a frivolous matter, fun is a social process that binds groups together in society. Though fun and frivolity can be partners, frivolity is better understood as subject-specific silliness that does not necessarily carrying the freight of meaning-making, whereas fun is much more functional. Fun has, in fact, been shown to be a foundational aspect of collective life that can create opportunities (and include commitment mechanisms for) communal identification, affiliation, and social critique.[6] Social scientists Gary Fine and Ugo Corte have provided more clarity around the meaning of fun through explorations of group interactions. They describe fun as a frequently unscripted component of small group cohesion and continuation that can emerge through group-recognized shared enjoyment and/or recollection of a shared past.[7] This framework is grounded in social psychologist Stephen Lyng's theory that fun belongs to a cluster of situation-emergent concepts that are emotionally fulfilling and freely chosen – not unlike the related concepts of humor, play, teasing, adventure, and games.[8]

This series of clustered concepts call to mind leisure (free time) pursuits of recreation, which dictionaries describe as "refreshment of strength and spirits after work" or "activity done for enjoyment when one is not working."[9] These ideas of *recreation* and *leisure* are often part and parcel of pejorative references to "fun" in Western culture, which has a long-standing tradition of conferring higher value on work and formal learning than refreshment or enjoyment.

---

[4] Podilchak 1991; Fine and Corte 2017.   [5] Beckman 2014.   [6] Ibid.
[7] Fine and Corte 2017.   [8] Lyng 2004.
[9] www.merriam-webster.com/dictionary/recreation;   https://en.oxforddictionaries.com/definition/recreation.

One of the most influential texts in the early twentieth century, Max Weber's *Protestant Ethic and the Spirit of Capitalism* highlights morally laden capitalist narratives that gained traction in the post-Reformation era, situating time as a commodity and disparaging non-income-generating efforts as theft from humans' highest purpose, the generation of wealth. Sociologists Luc Boltanski and Eve Chiapello have shown that capitalist discourses so consistently prioritized "work" and "work ethic" that primary affiliative relationships with occupation came to be a superordinate social identity in the twentieth century, weakening the importance of religious and civic affinity group identities in the shaping of dominant Western cultural narratives, and diminishing leisure as nonrelevant or frivolous.[10]

Though an alternative perspective gained some traction in 1936, shortly after Weber's treatise was published in English, when British economist John Maynard Keynes envisioned unprogrammed "purposeful leisure" as the pathway to higher-order mental activity,[11] Calvinist values remained entrenched. Social narratives of the past many decades accordingly framed zoos, as leisure destinations, cultural institutions of inherently low value, given that zoogoing appears to be entirely unrelated to zoogoers' accumulation of wealth. While recent studies have sought to identify how, and the extent to which, wealth *is* generated by museum operations,[12] for the purposes at hand we seek only to note a gap that has largely escaped notice. Capitalist economic narratives have tended to dissuade scholars and activists alike from exploring the structural importance of pleasure, enjoyment, fun, leisure, and recreation in the development of a sustainability-seeking, biodiversity-protecting society. Yet, happiness – an emotion often tied to pleasurable, enjoyable, fun leisure and recreation experiences – may be an inextricably useful role of zoos in adding social value through impacts on mental process and the development of collective environmental ethics.

Positive Psychology and the Pursuit of Meaningful Experiences in the Zoo

Positive psychology emerged as a domain of study in the 1990s after Abraham Maslow and others pointed out that psychologists had amassed

---

[10] Boltanski and Chiapello 2006.   [11] Weber 1905/2003; Keynes 1936.

[12] The think tank with which both authors are affiliated, Knology, has, for example, developed a series of publications on the regional and national economic impacts of children's museums in the United States. These reports were designed for members of the Association of Children's Museums and can be requested from the Association (see www.childrensmuseums.org/members/publications.

a significant body of research and applied practice around mitigating harms but had not given the same level of attention to the positive aspects of human thriving. Martin Seligman became a pioneer in the field of positive psychology when he began using the scientific method to develop a systematic theory about *human happiness* that was informed by his previous study of patterns of helplessness in dogs. Seligman describes happiness as positive emotion that can be actively pursued.

Awareness of and gratitude for the basic pleasures of companionship, the natural environment, and meeting our own basic human needs are the core components of a *pleasant life*, according to Seligman. *Good life*, he says, describes the deeper happiness achieved when we identify, cultivate, and make use of personal strengths and virtues in creative or practical ways. These dimensions of the pleasant and good life, though, are primarily embedded within the self. It is only when one begins discovering paths for consistently engaging uniquely developed personal strengths and virtues to achieve a purpose greater than the self – realization of a *meaningful life* – that an overall pattern of happiness might be expected, Seligman's data suggest.[13]

So how might the pursuit and realization of a pleasant, good, or meaningful life, thus conceptualized, be a valuable motivation and/or aspect of learning in the context of a zoo visit? With optimism researcher Christopher Peterson, Seligman identified six virtues considered attainable and inherently valuable personal attributes in almost every culture (see Table 5.1). Unlike *talents*, which Seligman defines as specific skills some (but not all) segments of a population naturally possess and can strengthen, *character virtues* can be learned and developed by most anyone through conscious personal effort. We note that every descriptor connected to the absence, excess, opposite, or realization (strength) of the character virtues that Peterson claims have direct impacts on human happiness sometimes play out as observable attitudes and behaviors when visitors navigate the social and value-laden experiences of a zoo visit – and can also overlap with the motivations and priorities driving and framing that visit.

John Fraser recalls, for example, a very long interview he once conducted with a family of extremely modest means who had moved to New York to seek economic opportunity after generations of their forebears had struggled through hardship as low-wage Mexican farm laborers. Now living near the Bronx Zoo, they tried to describe why they prioritized visiting that institution during the few hours every month when admission

---

[13] Seligman 2002.

Table 5.1 *Martin Seligman's summary of Christopher Peterson's list of universal character virtues and strengths*[a]

Character Virtue	Strength	Opposite	Absence	Excess
Wisdom & knowledge	creativity	triteness	conformity	eccentricity
	curiosity	boredom	disinterest	nosiness
	judgment	gullibility	ineffectiveness	cynicism
	love of learning	orthodoxy	complacency	know-it-all-ism
	perspective	foolishness	shallowness	ivory tower
Courage	bravery	cowardice	fright	foolhardiness
	persistence	helplessness	laziness	obsessiveness
	authenticity	deceit	phoniness	righteousness
	vitality	lifelessness	restraint	hyperactivity
Love	intimacy	loneliness	isolation/ autism	emotional promiscuity
	kindness	cruelty	indifference	intrusiveness
	social intelligence	self-deception	obtuseness	psycho-babbling
Justice	citizenship	narcissism	selfishness	chauvinism
	fairness	prejudice	partisanship	detachment
	leadership	sabotage	compliance	despotism
Temperance	forgiveness	vengefulness	mercilessness	permissiveness
	humility	arrogance	footless self-esteem	self-deprecation
	prudence	recklessness	sensation seeking	prudishness
	self-regulation	impulsivity	self-indulgence	inhibition
Transcendence	awe	criticism	oblivion	snobbery
	gratitude	entitlement	rudeness	ingratiation
	hope	despair	present orientation	Pollyannaism
	humor	dourness	humorlessness	buffoonery
	spirituality	alienation	anomie	fanaticism

[a] Transcribed by John Fraser from a summary PowerPoint presentation delivered by Martin Seligman at the 3rd World Congress on Positive Psychology in Los Angeles, California in June 2013, the presentation reflected on the scholarly work and life of Christopher Peterson. The table represents Dr. Seligman's efforts to honor Dr. Peterson's commitment to understanding and articulating positive psychology. Seligman described these character virtue and strength descriptors as a psychological framework for understanding the characteristics of living well, rather than a medical model anchored in a biological function. Seligman noted that this proposition is not fully representative of the emerging field of positive psychology but does represent Peterson's valuable contributions, which other scholars continue to build upon.

was waived. The family matriarch described the importance of learning from animals, claiming that "knowing" animals is a solid path to the development of a virtuous character. As her youngest son translated her Spanish, he noted his own surprise to hear the number of character virtues she associated with learning from animals. She spoke of using zoo visits to connect her children to wonderous life processes she herself had observed on the family farm, as well as opportunities to learn kindness, experience awe, and develop empathy for others. She matter-of-factly described these experiential learning outcomes as a crucial spiritual pursuit that would protect her children from risks they faced in their community, where violence, discrimination, and a rampant drug epidemic raged just outside their doors. While fears about these harmful influences led her to send her children to their grandparents' farm in Mexico every summer break, winters were marked by frequent decisions to visit the zoo. For this zoogoer, family zoo visits were happy experiences that (re)connected her family with nature and allowed for their spiritual grounding and growth.

This matriarch's perspectives, which point to how the learning and meaning-making that occurs in zoo settings tends to be simultaneously self-directed, personal, and co-constructed, also suggest that it is important for zoo staff and exhibit and program designers to recognize that pathways to learning and the seeking and development of personal strengths and character virtues tend to converge, why this matters, and in what conditions such developments may occur.

Like Seligman and Peterson, Mihaly Csikszentmihalyi, another pioneer of the scientific study of happiness, notes that while pleasure can be passive, *happiness is a state of being that must be prepared for and cultivated by using one's capacities with purpose.* Csikszentmihalyi found that intense focus to create or make sense of something perceived to be difficult but worthwhile can temporarily suspend the creator's self-awareness to such an extent that no competing thoughts, sense of time, or self-concern remain. Csikszentmihalyi's research has consistently shown that the merging of meaningful action and awareness produce this type of "flow," which becomes its own reward.[14]

While it is easy to imagine how the conditions for flow can arise through video gaming, or through the committed personal endeavors of a passionate poet, an elite athlete, an innovator, or a gifted musician, Csikszentmihalyi suggests that connecting deeply with another person or people can also be "flow activities," flow-producing conditions that make it

---

[14] Csikszentmihalyi 1990.

more likely a flow experience will occur. He notes that, like reading, connecting with others can be a mindful challenge because interpersonal experiences require complex skills of imagination and interpretation – and often nudge people just beyond their comfort zones. Such conditions facilitate learning and strengths development and can lead to the suspension of self-awareness as cohesive meaning-making occurs, according to Csikszentmihalyi, whose studies also suggest that resultant new understandings may continue to evolve through subsequent memories of the experience. Given the conditions now understood to support human strivings to achieve flow, it is perhaps unsurprising that cultural institutions have been shown to equip visitors with very diverse life experiences for flow experiences by provoking them to clarify and interpret concepts; weave new information into existing ideas and frameworks (of self, others, other species, our shared world, the spiritual domain, etc.); and reflect on potential implications. Indeed, Csikszentmihalyi points out that such outcomes are particularly likely when the unique goals motivating an individual converge with active, thought-provoking experiences that include opportunities for reflection.[15]

While conducting interviews asking adult visitors about their motivation for visiting the Central Park Zoo, in the heart of New York City, John Fraser and his team encountered one memorable solo visitor seeking just such an outcome. This middle-aged business suit–wearing man, who identified himself as an executive at the United Nations, said he made a weekly thirty-minute walk to the zoo to spend half an hour therein reflecting and "resetting his nature barometer." The seeking of this type of reflective pursuit and improved mental clarity is consistent with the "Spiritual Pilgrim" motivation identified as one of the five core zoogoing motivation groupings among zoogoers across the United States who participated in the WZAM study. More common in aquariums than in zoos, Spiritual Pilgrim visitors arrive with very distinct goals of reflection. Generally aversive to crowds, this zoogoing cohort seeks intimate, preferably unobstructed experiences and prefers program offerings at quieter times of the day or year.[16]

Seeking behaviors and the characteristics of flow are well known to the creators of "addictive" video games, which are very intentionally designed to become more complex and challenging over time, with increased skill

---

[15] Csikszentmihalyi 1997.
[16] 4 percent of the WZAM national study sample clustered around this identity-related motivation for their zoo visit. Falk et al. 2008.

demonstration eliciting immediate feedback. Beyond perceiving such gaming as enjoyable, players of all ages regularly find that their actions and awareness become integrated as they grow completely absorbed in the process of striving to reach unfamiliar levels of such games. This formula holds great appeal for educators, especially science-focused educators (within and beyond zoos), who recognize the potential benefits of "edutainment" – engaging and stimulating experiences designed to accelerate science-related or other specific learning outcomes by offering optimal conditions for flow (or at least a pleasurable dimension), including cognitive and sensory stimulation and immediate positive feedback.

Connections between zoos and immediate gratification call to mind philosopher Ralph Acampora's charged indictment of zoos as sites that appeal to hedonistic human tendencies, wherein the pleasure zoogoers experience through the exposure of rare and precious aspects of our biosphere for consumptive pursuits maps most closely to illicit, selfish, sexual gratification.[17] Leaders of the environmental movement, on the other hand, have overwhelmingly and consistently promoted frameworks and actions of constraint, self-denial, and delayed gratification, presenting a rationale for eudaimonic motivations and pleasures as more important, fulfilling, and responsible than short-term hedonic gratification.[18] However, we cannot discount the fact that leisure activities are, in and of themselves, inseparable from the sensations of the experience and how we evaluate and recall the experience.

While entertainment is no longer the foremost purpose of accredited zoos, many visitors do describe recreation as an important motivation for their visit,[19] and recent data suggest that most zoogoers claim their visit made them happy.[20] As mentioned in Chapter 4, though, zoos' efforts to balance entertainment and marketing strategies with conservation education mission goals are often criticized.[21] Claiming, for example, that zoos are unable to effectively and appropriately educate public audiences about conservation, Keekok Lee opines zoos' legitimacy as civic actors rests only in the success of these institutions as venues for "wholesome family recreation and entertainment."[22] Yet, researchers have now confirmed that the pursuit of recreation through a zoo visit can support and advance zoos' education and conservation goals.[23] We will thus shift to unpacking what social science researchers have explored and learned about the value of the

---

[17] Acampora 2001, 2005, 2006.   [18] Hamilton et al. 2018.
[19] Kreger et al. 1998; Sickler and Fraser 2009.   [20] Clayton and Le Nguyen 2018.
[21] See, for example, Mason 2000, Acampora 2001, 2005, 2006, Beardsworth and Bryman 2001, and Packer and Ballantyne 2004.
[22] Lee 2005.   [23] Kreger et al. 1998; Fraser et al. 2008; Sickler and Fraser 2009.

zoo as an educational leisure venue in order to provide context for an assessment of what pathways to flow might be facilitated to impact conservation learning experiences and outcomes in zoo settings.

Understanding the Zoo as an Educational Leisure Venue

Based on a thorough review of the visitor motivations and satisfaction literature, Jessica Sickler and John Fraser concluded in 2009 that the majority of studies related to visitors' decisions to spend leisure time at the zoo used survey methodologies to explore attendance patterns, zoo-going motivations, or expectations of and satisfaction with specific amenities and tangible services visitors might have experienced during their visit.[24] These researchers noted that much of the existing research around perceptions of zoos, motivations for visiting, and on-site experiences prior to 2010 should be reviewed with caution because research tools have often lacked clarity regarding which aspects of a zoo visit might be connected to recreation, enjoyment, and learning – or how such concepts should be understood.

In the early 1990s, for example, Kathleen Andereck and Linda Caldwell developed a metric that asked visitors to describe their motives for visiting the zoo and rate their level of "general satisfaction" with respect to educational aspects of the visit, recreational aspects of the visit, optimal arousal, and perception of crowding. While these authors found that recreational and educational aspects of the visit were the strongest motivating factors for respondents and the strongest predictors of respondents' satisfaction, the survey instrument did not define education, so it is difficult to know what respondents assumed about this question.[25] Concept ambiguity may also have been a factor in a study shortly thereafter in the United Kingdom finding that respondents described conservation as zoos' highest priority and entertainment as zoos' lowest priority, yet the metric used for analysis demonstrated that the primary motivation for these same respondents' zoo visits was entertainment.[26] In 2004, a

---

[24] In his 2006 literature review of zoo audience research, Davey distinguished visitor information collected by marketing professionals from behavioral observations undertaken by academics and zoo staff, an arbitrary distinction that ignored possible overlaps as well as whether the constructs used in traditional marketing analysis are appropriate to zoo settings. Furthermore, the studies reviewed by Davey did not probe how education experiences were defined or understood by study participants, which limits the usefulness of these findings and associated assumptions about the contribution of public amenities to overall zoo popularity or the mission impact of zoo experiences.

[25] Andereck and Caldwell 1994.   [26] Reade and Waran 1996.

research study developed to distinguish "benefits" from "services" in order to clarify the contribution of education experiences to overall expectations about zoo visits showed rather perplexing cognitive dissonance, in that zoogoers desired wildlife information but did not identify "education" as a high priority.[27]

Throughout the world, zoo visits rank as one of the most popular leisure pursuits for families,[28] yet "family-friendly" has proven to be another elusive construct in the realm of research. Tomas, Crompton, and Scott found that many respondents in their sample valued zoos as "family-friendly" venues, for example, but did not probe why respondents characterized zoo venues this way.[29] A more refined definition of "family-friendly experience" emerged from Sophie Turley's 2001 study showing that adult respondents in the United Kingdom primarily characterized zoos as venues that facilitate bonding experiences and shared pleasurable activities with children, rather than as venues with an educational purpose. While respondents did specifically attribute "learning about animals" as a family-friendly aspect of zoo experiences, study constructs did not define or explore the perceived value or dimensions of learning associated with this attribute.[30]

Problematically, in these and many other studies, concepts of recreation/entertainment and enjoyment/satisfaction have often been presented as motivations and outcomes distinct from, or in contrast to, education/learning. That is, few researchers to date have included recreation/entertainment or education as potential variables *of* enjoyment/satisfaction in the assessment tools they use to measure and understand the attitudes, perceptions, and experiences of zoogoers.

Hoping to clarify the complex relationship between entertainment and education through a study of learning as a leisure pursuit, Australian researchers Jan Packer and Roy Ballantyne engaged visitors at a range of educational leisure settings in extensive qualitative interviews prior to their development of a quantitative assessment metric. Their research, published in 2004, shows that the content of an attraction is partially independent of the pleasure derived from the process of discovering new connections and relationships – demonstrating that some visitors identify their personal learning process as entertaining, pleasurable, and satisfying. Importantly, Packer and Ballantyne were among the first researchers to situate learning

---

[27] Ryan and Saward 2004. Marvin and Mullan 1987 speak to this pattern, as well.
[28] Baratay and Hardouin-Fugier 2002; Fraser et al. 2008.   [29] Tomas et al. 2003.
[30] Turley 2001.

as a leisure pursuit personally conceptualized by each visitor, rather than focus on measuring whether visitors had acquired specific knowledge prioritized by the venue.[31]

Fun and Meaning in the Zoo: Education Psychology Research

Research in zoos and other types of museum has confirmed that education, fun, social interaction, and enjoyment should be understood as aspects of the visitor experience that tend to blur, overlap, and reflect significant individual variation. Jan Packer's education psychology research has shown, for example, that visitors who enjoy expanding their existing relationships, solving puzzles, or encountering new structures for organizing existing knowledge often perceive learning within cultural institutional settings as a pleasurable experience – but not all visitors value such endeavors or find such experiences fun.[32]

These realities make it extremely challenging for zoos to identify what experiences make learning in zoos fun and meaningful to different people.[33] Even the animals, the distinguishing institutional feature everyone cites when describing why they visit the zoo, are often unpredictable – and different people find different zoo animals fun and meaningful for many different reasons.[34]

WZAM2 research included an enjoyment study that sought to get to the heart of subjective perspectives about zoo experiences adults (the primary visitation decision-makers) identify as fun, in order to examine whether varied perspectives might have some features of coherence that could inform zoo programming and evaluation. Aesthetic appreciation of animal beauty emerged as a rather strong contributing factor to enjoyment for all zoo visitors in the WZAM2 sample, whereas the idea of enjoying the zoo as a getaway from the day-to-day routine emerged as neutral. In addition, four distinct themes arose with respect to personal perceptions about what makes a zoo experience enjoyable, with different combinations

---

[31] Packer and Ballantyne 2004.   [32] Packer 2006.

[33] In 2008, David Klenosky and Carol Saunders endeavored to explore zoo visitors' perceptions by connecting their descriptions of motivations and satisfaction to specific aspects of their experiences in the zoo. These researchers found that enjoying the visit and learning about animals and nature were overarching motivations associated with deeper values; their data pointed to various ways zoo attributes can contribute to visitor satisfaction but did not indicate how multiple experiences interact to impact a visitor's perception of overall enjoyment from the visit.

[34] Falk et al. 2008; Fraser et al. 2008.

of psychological and sociological value areas impacting visitors' perceptions of overall enjoyment.[35]

Perspective A respondents, those who identified the social experience within the family group as the defining factor of enjoyment, did not report that cognitive stimulation (the process of discovering or acquiring new information) contributed to their enjoyment at the zoo but did place high value on seeing animals, especially if the experience was close up or involved seeing them move and play. Those with this perspective did not report feeling any sense of connection with the zoo's conservation mission or the animals themselves.

Perspective B respondents were also largely focused on social family experiences as the dominant factor of enjoyment at the zoo, but these respondents described family experiences that included cognitive stimulation, peace and tranquility, and a sense of connection to the zoo's conservation mission as fun. These respondents valued the outdoor setting of the zoo, but, unlike Perspective A respondents, derived particularly little enjoyment from sensory experiences related to zoo animals or being entertained by zoo animals, and strongly rejected interaction between humans and animals through exhibits or demonstrations of animal training as enjoyable aspects of their visit.

In great contrast to Perspectives A and B, Perspective C respondents emphasized seeing zoo animals as their primary source of personal enjoyment, and social aspects of the zoo visit either did not contribute to, or even detracted from, their overall experience. Emotional responses and feeling connected to the animals on display was the most common source of animal-related enjoyment for these respondents, and even cognitive stimulation, which these respondents identified as strongly contributing to enjoyment, was intertwined with the experience of seeing the animals. Though they reported that social interactions did not contribute to their enjoyment of the zoo overall, most Perspective C respondents had come to the zoo with their families or as members of a group.

The few Perspective D respondents distinctly defined enjoyment through social experience, but with friends, rather than family. Highly interested in learning about their (adult) companions, these respondents

---

[35] For a thorough description of the Q methodology used for the WZAM enjoyment study, see Sickler and Fraser 2009, 316–321. The study was designed to look at the mix of zoogoers' perception of enjoyment through their layered experiences overall, rather than in isolation. Fun and enjoyment were deliberate foci, as constructs of satisfaction or motivation that had been addressed in previous research and these ideas more directly represent the language zoo visitors use to describe their perspectives of on-site experiences.

valued the self-directed roaming and exploration of the large variety of exhibits and activities in the zoo. Sensation-seeking experiences were much more strongly correlated with enjoyment for these respondents, particularly experiences of excitement and fear, or seeing animal–visitor interactions. Child-focused and traditional elements of zoo programming such as shows, animal training demonstrations, and getting information from zoo staff were not of interest to Perspective D respondents, who define their relationship to the zoo on their own terms.

This area of research shows that the reasons the public values the zoo are not important or distinct from learning, and the importance of identifying and examining coherent patterns related to how visitors construct enjoyment from their own point of view. Previous researchers have shown that in-group social interactions and experiences with animals consistently emerge as defining zoogoers' motivations or satisfaction, but the WZAM findings point to the different interplay patterns of these value areas reflected in the enjoyment narratives of each individual.[36] These studies support the conclusion that institutions focused on visitor enjoyment and learning must go beyond demographic visitor segmentation and explore how visitors to their sites perceive and process their experiences therein, as even visitors within the same visiting group often have divergent perspectives regarding what about a zoo visit contributes to enjoyment.[37]

Whenever an individual's motivations and expectations align with their experiences, the data suggest that individuals will mentally categorize the overall zoo experience as positive. Some common expectations of visitors seem to be well served by the current structure of zoo exhibits and messaging. Zoos already focus, for example, on evoking the personal emotions many visitors experience and value when they see and smell and hear live animals (Perspectives A and C). Adult visitors who derive significant enjoyment from children's experiences and excitement, on the other hand (Perspectives A and B), are likely to be more responsive to conservation-messaging strategies that use a social perspective: an under-developed opportunity for zoos. Furthermore, though zoogoers focused on introspective experiences observing animals (Perspective C) represent the visitor segment most favored by conventional exhibit design and curation approaches that situate visitors as learners lacking social context, Perspective C respondents represented a small segment of the overall

[36] See, for example, Andereck and Caldwell 1994, Turley 2001, Tomas et al. 2003, and Ryan and Saward 2004.
[37] See, for example, Morgan and Hodgkinson 1999 and Davey 2006.

sample.[38] While zoo staff may be discouraged that only visitors who arrive with Perspectives B and C focus explicitly on learning as a component of enjoyment, and perhaps more so that only Perspective B visitors report zoo mission as impacting their enjoyment, visitors may learn more (and come to care more) about conservation due to their visit, even if zoos' conservation mission does not inspire their enjoyment per se.

Importantly, the perceived value of zoos as leisure experiences appears to be tempered by visitors' moral evaluation of how captive animals are being cared for. Barbara Woods' investigation of the impact of zoo experiences on visitors' perceptions of animals highlighted aspects of learning associated only with zoo visits (not visits to wild settings). Her findings also show that visitors' perceptions of the quality of animal care at the zoo have a direct impact on their perceptions of enjoyment, as well as the value they attribute to learning in the zoo setting.[39] Other research similarly suggests that animal care is a mitigating factor in visitor satisfaction at zoos.[40] Such findings suggest that zoo satisfaction and learning studies that have not asked about visitors' perceptions of animal care probably reflect incomplete or misleading conclusions about how visitors experience and value zoos. Future investigations of educational leisure settings can build on these findings by defining children's perspectives of enjoyment in zoos and further exploring the relationships between visitors' personal value areas and the social experiences, metacognitive processes, and identity development that occurs in zoo settings.

Emotional Response Patterns and Educational Leisure Learning in Zoo Settings

Educational leisure fits easily into concepts of learning as a lifelong pursuit. Zoos appear to draw casual visitors to engage in discoveries of personally relevant information, though many visitors may not characterize such discoveries as "learning." There are certainly many opportunities to move beyond one's comfort zone, learn, and develop personal strengths and character virtues during a zoo visit. Even if learning is not expressed as a priority or motivation for zoo visitors, many of the studies described in this chapter show that zoo visitors do consistently consider learning to be a

---

[38] For examples of such exhibition design narratives, see Coe 1985, Polakowski 1989, and Pekarik 2004.
[39] Woods 2002.   [40] See, for example, Ryan and Saward 2004.

valuable part of their zoo experiences. So, what pathways to learning might be optimal for advancing a conservation agenda in zoo settings?

Humans frequently insist that they respond to their environment and learn with a rational mind, yet a robust body of neuroscience data shows that emotional arousal and response patterns play an important role in the learning process. A growing number of neuroscience researchers suggest that the affective domain must be understood as a crucial mediator of learning because deep emotional attachment to a subject area or concept facilitates deeper connections to and understandings of related information.

In the 2010s, there was substantial effort to explore the affective dimension of zoogoing. Jerry Luebke, a researcher at Brookfield Zoo in Chicago, collaborated on a series of studies demonstrating that the pleasure of viewing animals is emotionally fulfilling and helps zoo users make meaning of their experiences in ways that align with mission goals. Using mixed methods over a number of years, Luebke and colleagues have been able to explore and articulate the degree to which the pleasurable aspects of viewing live animals seem to fulfill not only a desire to see animals, but also a desire to make direct connections regarding why and how conservation might be valuable for the protection of various species. These researchers have demonstrated that positive affective experiences result in a heightened level of empathy for animals, directly predictive of visitors' sense of connection and desire to see increased conservation effort to protect those species.[41]

Niels Bonderup Dohn recently used a case study approach to identify sources of interest and attributes that stimulated interest when Grade 12 biology students in Denmark took a field trip to a zoo. Using a descriptive interpretive approach based on classroom and field trip observation, video data, and interviews, Dohn found that affective experiences can stimulate students' learning motivation during a group zoo visit if triggered by such variables as hands-on activities, novelty, surprise, knowledge acquisition (through activities and information transfer), and opportunities for socialization – and strong interest was stimulated when several variables occurred simultaneously.[42] Consistent with the range of value areas and enjoyment perspectives identified through the WZAM enjoyment study, these findings point to the need for further study of affective experiences that may stimulate learning motivation at the zoo.

Neuroscientist and psychobiologist Jaak Panksepp, pioneer of the interdisciplinary field of affective neuroscience, analyzes the autonomic

---

[41] Luebke and Matiasek 2013; Matiasek and Luebke 2014; Luebke et al. 2016; Luebke 2018.
[42] Dohn 2013.

functioning patterns of humans and related animals. Panksepp has identified seven automatic emotional arousal systems that provoke predictably
similar responses in humans, other primates, and even cats and rats:
seeking, play, lust, fear, panic/grief, rage, and *care.*[43] Perhaps because the
zoo uniquely connects humans with comparatively unfamiliar animals in
social settings that include numerous sensory stimuli and elements of
unpredictability, zoogoers tend to experience several of these emotional
sensations while on site. Here, we briefly reflect on how specific emotional
response patterns commonly manifest in the zoo, sometimes in ways that
appear to be connected to zoo visitors' overall experiences and learning.

Seeking and Play

While design stories in zoo exhibits reflect the pedagogy of the zoo, zoos are
also set up as a puzzle piecing together sociocultural stories of human
relationships to nature, place, etc. "Who and where am I?" can thus become
a seeking question that provokes visitor engagement. Recognizing that
explorative behavior is a pathway to learning, all museums are intentional
about bringing together objects and ideas that feed humans' seeking behaviors through material culture. Seeking, we note, has smooth alignment with
the "Explorer" identity-related motivation for visiting the zoo that characterized a significant segment of WZAM survey respondents.

Play, the action of exploring contingencies to understand how a system
works and make sense of things or situations, is connected to mammalian
social principles that cement social-support networks and friendships,
according to Panksepp.[44] Famously described by Jean Piaget as "the work
of childhood," play is often counterposed to wisdom and knowledge and
presented as an obstacle to learning. In his landmark work "The
Importance of Play," Bruno Bettelheim attempts to describe the psychological importance of play on the developing minds of young children.
According to Bettelheim, play is a nonfrivolous intellectual activity because
as they play-act, role-play, and use their imagination, children develop
cognitive skills (sorting, basic math, etc.) and social skills (negotiating, etc.)
they aren't even aware of learning, an outcome that brings us back to
human pursuits and achievement of flow.[45]

---

[43] Panksepp 2011.
[44] Ibid., 10. Care, which will be discussed further in Chapter 7, is an instinct that frequently overlaps
    with play – and supports similar outcomes in a range of mammalian species, according to Panksepp.
[45] Bettelheim 1987.

Dutch cultural historian Johan Huizinga made a compelling case for the "civilizing function" of play as a core element of culture and society, an instinct that is present in law and in war, as well as in art.[46] Huizinga notes that while enjoyment may be a motivation for play, play often becomes serious behavior that "creates order." Indeed, chess and checkers are war strategy games that give players opportunities to navigate offensive and defensive contingencies, and these games are widely enjoyed by adults and children alike. John Fraser notes that fathers visiting the zoo with their children frequently act brave in contexts of playful discovery when, for example, the "dangerous" lion is close by (but, of course, very safely confined to its enclosure). So, in addition to being intentionally facilitated by various intentional aspects of zoo design, seeking and free and guided play are learning-supportive endeavors and experiences that arise in the zoo when personal motivations/priorities/pursuits, on-site stimuli, and inter-personal experiences intersect, as well.

Lust

Lust is an automatic emotional arousal system that is particularly interesting in zoo contexts because zoos are highly sexual environments. Studies of pheromone concentrations in zoo animals such as elephants have shown that the atmosphere of the zoo environment itself is, literally, chemically, sexually charged.[47]

Sensory pleasure centers in the brain tend to become activated in the zoo by variations in exercise, visual stimuli, experiences with water, and smell. While visuals tend to be the focus, scents immediately enter our brains and trigger memories and emotional responses. Olfactory communication, furthermore, regulates sexual attraction and social behaviors for many species. Unlike odors, which are molecules that activate areas of the cortex and nervous system (areas associated with cognitive processes of learning and memory), semiochemicals are signaling molecules that do not become synonymous with past experiences.[48] Many zoos periodically disperse specific semiochemicals animals use for communication because they have been shown to reduce undesirable behaviors such as aggression, fear, stress, and boredom in a wide range of animals with no undesirable side effects, and appear to increase play and other affiliative social behavior in captive lions.[49] Recent studies indicate

---

[46] Huizinga 1938.   [47] Rasmussen et al. 1997.   [48] Tirindelli et al. 2009.
[49] Martínez-Macipe et al. 2015. The captive lion study took place in only one zoo with eighteen lions, so additional studies will be needed to confirm the researchers' findings.

that humans also use olfactory communication and produce, perceive, and respond to species-specific semiochemical compounds (pheromones).[50]

So, human primates also experience a pheromone information bath when *we* spend time in the zoo, though we are undoubtedly unaware of the many ways the odors and semiochemical compounds in those settings impact our overall experiences. Perhaps, though, these realities partially explain why zoos are a popular place to get engaged, and the national WZAM study finding that 13 percent of non-solo zoos and aquarium visitors in the sample arrived with a date.[51]

We do know for sure that prurient curiosity frequently arises in zoo settings. The sexual and reproductive behavioral norms of animals and the biological function of different species are undoubtedly important aspects of conservation, too, yet zoogoers, at least in the United States, tend to approach topics of sexuality and reproduction through humor because they are uncomfortable discussing these matters.

John Fraser was part of a Bronx Zoo team that created videos of tigers' play, predation, and sex behaviors. While adults often share knowing smirks or become visibly uncomfortable when they encounter bears or otters engaging in self-pleasure within a zoo exhibit, numerous parents reported finding that the sexual content in the tiger videos was welcome and appropriate because they felt the zoo was an ideal context for conversations with their children about realities of biology and socialization. Despite Fraser's observation that most zoogoers make erroneous heteronormative assumptions around sexually dimorphic animals and dominance hierarchies on display at the zoo, heterosexual *and* homosexual Valentine's Day programming at many zoos in Western countries reflects these institutions' awareness that they offer opportunities for diverse audiences to navigate and make sense of "natural" sexual and social norms.

Fear and Panic

Touching back to olfactory experiences and involuntary emotional responses in the zoo, animals and humans alike can detect fear as a smell. While phobias and aversions to specific taxa, species, and natural conditions have contributed to human fitness and survival over time, such responses can also deeply impact how some people experience nature.[52] For some people, aversion can be strong enough to ground a kind of fascination and striving to understand their fear and demonstrate that they

[50] Grammar et al. 2005.   [51] Falk et al. 2008.   [52] Kellert 2012.

are in control by, for example, intentionally seeking out a snake, bat, or spider zoo exhibit.[53] Aversion can also lead people to characterize some forms of life as having lesser worth.[54] Many zoos now offer touch tanks and live animal presentations that demonstrate how zoo staff care for species like spiders, snakes, cockroaches, sharks, rays, and other species believed or known to threaten human life. Such programming seeks to communicate – and allow audiences opportunities to reimagine and negotiate for themselves – reasonable fear and what is "known" about aversive animals, giving public audiences structured opportunities to tap into, discuss, and hopefully learn from/through, their own natural emotional systems of fear and panic in a setting perceived to be safe.

Animal interactions have long been used to address fear and studied as an undertaking that can help children develop agency,[55] and researchers in the twenty-first century have shown that adults and youth alike report less fear of animals after exposure to related educational programming.[56] Yet while programs or presentations at zoos and aquaria that incorporate a live animal have been shown to significantly increase both engagement time and knowledge in guests,[57] short- and long-term attitudinal and behavioral outcomes of such interactions remain understudied.[58]

Serious study of the social psychology, sociology, and biochemical domains as an integrated area of exploration is likely to be relevant to understandings of learning pathways in zoo settings. Such a lens is appropriate for the study of, for example, what happens when public audiences experience up-close encounters with aversive "ambassador animals" – in a social context characterized by a lot of olfactory stimuli, seeking, and play. Do emotions such as fear and panic that sometimes emerge in these situations appear to facilitate learning? Can such experiences increase participants' interest in (and perhaps concern for) uncharismatic or aversive species? Inspire these individuals to support or even motivate them to act for the protection and well-being of that and/or other species in the wild?

To begin to ascertain whether ambassador animal programming does, indeed, appear to help advance zoos' conservation mission, members of the AZA's interdisciplinary Ambassador Animal Scientific Advisory Group (AASAG) began to research how interactions with ambassador animals at the Bronx Zoo impact visitors' curiosity, connections with animals, and conservation intentions. In 2019, this team initiated a study using a semi-

---

[53] See, for example, Kahn et al. 2008.   [54] Kellert 1997; Herzog 2011.   [55] Myers 2007.
[56] Røskaft et al. 2003.   [57] Heinrich and Birney 1992.   [58] Kisiel et al. 2012.

structured quantitative observation protocol and grounded theory interview approach to gain insight into affective responses, learning indicators, and conservation action interest through twenty programs with a range of animals. Common themes highlighted by the data suggest that facilitators can encourage curiosity about animals through questions posed to participants about the ambassador animals' natural history. Inquiry-based facilitation can also help facilitators understand where their audience is, conceptually, and where they can go with content knowledge. While the researchers' observations of facilitated inquiry focused on natural history, there is reason to believe that inquiry can be leveraged to explore other topics with visitors. The study also showed that facilitators can foster connections between visitors and animals through close-proximity experiences, and by engaging visitors through stories about the individual animals. Visitors who heard about animals' individual habits seemed more likely to identify the animals as individuals. While the research team seeks to test new ways to inspire conservation-related actions in 2020, previous research suggests that strong ties between the animal/s and the actions, as well as immediate on-site action opportunities, can help turn conservation-related intentions into actions.

Conversation and engagement in scientific reasoning – important aspects of informal culturally framed meaning-making and science learning – has been shown to occur at touch tanks. According to a recent study conducted at four West Coast aquariums in the United States, families are making and challenging claims, and in some cases seeking confirmatory evidence, to make sense of new information and concepts in ways they find useful and meaningful.[59] Data revealed that, while talk about ecology was limited, conversations about ecology between visitors and zoo staff were significantly longer than ecology conversations between guests only, suggesting that facilitators can leverage tactile experiences to encourage mission-related discourse at touch tank exhibits.[60] Saint Louis Zoo researchers also studied the effects on visitor stress levels associated with interactive touch tank experiences at a pool featuring stingrays, sharks, and horseshoe crabs. Stress was measured in two ways: heart rate monitoring before, during, and after the experience; and mood assessment (using an established psychological instrument) before and after the experience. Findings suggest that interacting with the animals at this touch pool was physiologically akin to theme park experience outcome norms and led to a decrease in mental stress.

[59] Ibid.   [60] Kopczak et al. 2015.

Reptiles, amphibians, fish, and invertebrates seem vastly different from people, which may be why these types of animals have not traditionally been included in empathy studies related to the development of environmental concern.[61] Zoogoers smell, and also taste, guano; it's in the air around the bat exhibit. Yet, aversion can help practitioners introduce these species to public audiences in sensitive and responsible ways, even as they seek to use emotional response mechanisms as potential pathways for advancing a conservation agenda. Affective components of meeting animals up close, furthermore, can be more memorable than other types of exhibit experiences. While zoogoers approach their visits with different motivations and existing understandings, on-site educators/facilitators have an opportunity to leverage up-close experiences for teaching science content and concepts; sparking interest in animals; connecting visitors with animals; and guiding visitors to expand their conservation-related actions.

Learning Is *Pleasurable*

In summary, zoos and aquariums present their users with a wide range of novel puzzles to solve. Together, we discover new things about ourselves, each other, and the world around us when we spend time in a zoo. Numerous stimuli impact our every sense, and these are experiences we can have and share in public. The mental and emotional process of sense-making in such contexts is frequently a pleasurable pursuit. While that process and the overall array of experiences that provoked it can be described as fun, that descriptor is too often misconstrued as suggesting frivolity. While enjoyment experienced at the zoo may be hedonic or eudaimonic, neither precludes the idea that users' novel experiences at zoos and aquariums contribute to learning. Human thriving, an inherent attribute of a positive psychology concept, is a pursuit and outcome that people can explore collectively and within themselves – and this is an important affordance and benefit that individuals, groups, and societies can derive from the cultural institutions to which they have access.

---

[61] While animal welfare has become an emotionally charged and political issue for charismatic megafauna, little is known about nonmammalian species, a point that has also been raised by numerous scholars. See, for example, Myers et al. 2004, Sickler et al. 2006, Goulart et al. 2009, Melfi 2009, Burghardt 2013, and Horvath et al. 2013.

CHAPTER 6

Meaning
Constructing Knowledge through Discourse, Dialogue, and Metaphor

John Fraser once interviewed a sample of adults to learn more about formative life experiences at zoos when they were growing up. Then in his mid-thirties, Greg had long ago left the Midwestern suburb of his childhood and achieved early success as an Ivy League professor known for his highly respected conservation biology research leadership in the tropics. In the middle of his interview, Greg grew concerned that his fond memories visiting the zoo with his mother were somehow inaccurate because at times these representations seemed discordant with what he knew about his mother's values and beliefs. Greg abruptly excused himself, pulled out his cell phone and rang his mother to discuss their conversations and experiences together when he was a child, and explore her thoughts and memories. When they finished, Greg relayed the conversation:

> So, we confirmed it, though only after I said, "why did you take us to the zoo, what did you want me to get out of it?" She immediately went to, "we wanted you to experience, we wanted you to learn, wanted you to have another opportunity to see all of these things that I knew you were interested in." But when I prompted her with "some of it was you wanted me to develop respect for the animals," she said, "oh, well yes. Absolutely. I just didn't think to say that because it seemed so obvious." And she talked about sort of, modeling behaviors, that kind of thing. So, she said, "Yeah, that was definitely a part of it, that we went to sort of develop that connection." So, I was right when I said, "oh, absolutely, that was a reason," but, interestingly, it wasn't the first one she offered until I prompted it."

While Greg readily recalled his mother's efforts to nurture his knowledge acquisition and self-esteem as a scientific thinker, it was only after he was prompted to reconstruct his own memories of *how* his mother provided validation that he came to realize memorable experiences at the zoo were part of an ongoing dialogue directly connected to the values and beliefs his mother still sought to share with and instill in Greg and his siblings. Supported by a call to his mother to benchmark a new discovery, Greg came to realize that his childhood zoo experiences had not been transactional, but were part of a long-term, intergenerational negotiation of nature values, shared experience, and desirable behavior.

Data introduced in Chapters 3 and 5 revealed that people of all ages use an assortment of animals and people in parklike settings for their own learning purposes – frequently as a tool to deepen a narrative or conceptual feedback loop related to their motivation for visiting. They work to make sense of new or unexpected information or experiences. Though conservation scientists, zoo designers, and zoo staff aim to leave zoogoers with deeper factual knowledge, the learning paths visitors seek and pursue often involve aesthetic experiences, personal emotions, and moral framing that are much broader than the learning outcomes imagined in the development of messages by zoo staff. Greg and his mother show us how people construct knowledge by building on prior beliefs, values, and perspectives that continue to be shaped through action, interaction, and sensory experience.

Like meaning-making experiences in the zoo, threats to biodiversity are numerous, layered, and fluid, and people are involved and impacted in different ways and to different degrees. The same pattern applies to conservation action opportunities. This is, of course, why consensus about how to define, prioritize, and pursue conservation outcomes is so elusive.[1] This is also why, rather than seek to enforce a singular narrative, agenda, and strategy for conservation, we feel zoo leaders should focus on establishing and developing their spaces and their staff as flexible resources equipped to facilitate idea sharing and accountability through civic mechanisms of discourse development and dialogic exchange.

This chapter explores some of the discursive tools that zoo visitors use to inform, organize, and interpret their on-site experiences. We begin by introducing discourse analysis as a helpful tool for understanding how humans create meaning through and about their zoo experiences. We highlight dialogue and metaphor as distinctly relevant discursive mechanisms for the learning that tends to occur as zoo visitors negotiate multiple narratives and multisensory, emotional experiences when they encounter animals and people during their visit – and simultaneously and subsequently integrate those experiences into their existing mental, narrative, and moral frameworks.

---

[1] For a thorough assessment of the many ideological underpinnings grounding the perspectives and pursuits of individuals and groups claiming to speak and act for nature, see Darier's 1999 edited volume, *Discourses of the Environment*.

Discourse

Language is now widely recognized as a medium of negotiation interconnected with teaching and learning.[2] Embedded in language, *discourse* is a shared way of making sense of the world, facilitating shared interpretation and the development of coherent narratives. We define conservation discourses, for example, as discourse based on an ethic of human care and responsibility for nonhuman animals and for the ecological systems that support diverse forms of life on a shared planet. Such an ethic, and such discourse, gives shape to individual and collective practice of tempering materialist values with an ecological worldview and commitment to reducing or minimizing environmental harm.

Much like the range of zoo critiques and zoo value discourses highlighted in previous chapters, conservation-focused discourses take various – and frequently divergent – forms. These narratives reveal different conceptions about how best to address complex concerns and opportunities.[3] Divergent discourse directions can be seen, for example, in the written records of the New York Zoological Society. Unlike his predecessor, William Temple Hornaday, whose written and verbal communications focused much more narrowly on the role of elites and experts as policy influencers, Fairfield Osborn's writings and radio broadcasts reflect discourses grounded in the conviction that problem-solving discourse should emphasize the role of citizens as producers and consumers capable of collective democratic organizing. While both leaders recognized the social dimensions of ecological issues and were passionate about conservation, which they saw and pursued as an urgent undertaking, their discourses and the strategies associated with those discourses, were distinct and were effective in different ways.

The shared terms and assumptions that emerge or are reinforced through discourse are important first steps for the robust dialogue that is typically needed to solve complex problems involving more than one individual. An emerging body of research suggests that zoos can successfully nurture open dialogue and critical questioning about how established power dynamics, social practices, and personal decisions impact biodiversity and conservation. Here we take a deeper dive into the diverse

---

[2] Vygotsky 1978; Halliday 1993; Ash 2003.
[3] Interests and power are context factors that directly impact conservation and can also limit the influence of conservation discourse. Dryzek 1997 offers a comprehensive starting point to explore through a historical lens how context dynamics have impacted the strengths, weaknesses, and influence of well-established environmental discourses.

discourses zoogoers arrive with and develop on site, in order to ground further exploration of how zoo spaces and zoo staff might facilitate authentic dialogue that supports the development and adoption of strategies and commitment to advance conservation efforts and outcomes.

Discourse Analysis

Research in museum settings demonstrates that live animal displays elicit amongst visitors more diverse learning conversations than typically occur in science museums overall.[4] In particular, the opportunity to learn alongside others while safely in the presence of living animals appears to directly provoke shared discursive processes that are often connected with knowledge integration and group identity development.[5]

Trained originally as a zoologist, Dr. Sue Dale Tunnicliffe introduced educators and zoo staff alike to a novel data source when she began to record and assess what zoo visitors really talk about – sometimes by hiding and overhearing, using the same data collection strategies zoologists use to study nonhuman animals. An early advocate for using discourse analysis as a methodological tool for investigating intergenerational scientific meaning-making in museum settings, Tunnicliffe was among the first to confirm that comments about animal exhibits in museums and live animal displays in zoos are drawn from visitors' existing knowledge and expectations, rather than interpretation the institution provides.[6]

In fact, while animals often provide unedited experiences that spark dialogue among visitors, zoo*goers* sometimes become the topic of intense focus and dialogue. Many people who work in zoos, for example, share anecdotal experiences of visitors who have placed themselves at risk by attempting to breach an enclosure barrier designed to separate zoo animals from zoogoers – barriers that many zoo animals come to understand as landscape features that prevent threats and allow them to more or less ignore the public as they pursue their daily routines. Though uncommon, barrier-breaching efforts get people talking, particularly when the determined interloper removes clothing in the process, a quite unexpected happening that John Fraser notes occurs periodically at the Bronx Zoo. While zoo staff, visitors, and the general public alike might identify and describe such an event as an uncommon pathology, the explanation given by the nudist usually includes reference to the spiritual value of connecting with (comparatively) wild nature.

[4] Allen 2002.   [5] Myers 2007.   [6] Tunnicliffe 1995.

While we can't conclude that rule-breakers who want to get naked and hug animals at the zoo are enacting personal fantasies of being one with nature, or that those who attempt to sneak into an enclosure with large carnivores are suicidal, what is clear is that these individuals share the belief that the living animal/s they took such measures to approach have some form of relationship with them.

Dialogue

Psychologists have shown that individuals and groups build on and fine-tune their relational, moral, and conceptual frameworks when they share perspectives and worldviews. This is the reason dialogue – the exchange of ideas and opinions through listening and discussion that allows participants to confirm, contest, and create various discourses and narratives – has such a powerful impact on the constant evolution of language and culture.[7] While new perspectives and behavioral norms cannot be imposed on those who introduce, co-create, or listen to a conversation, dialogue can render an alternative or unfamiliar narrative more comprehensible, relevant, and/or legitimate.[8]

Researchers have begun to focus on the specific thematic content and categories of talk that underpin family dialogue about life sciences in museum settings.[9] As described in Chapter 3, zoo visitors who fit the profile of Facilitators can be understood as concept-framers who are very intentional about conservation topics they introduce and are themselves often passive learners. Fitting the profile of Facilitators, teachers visiting the zoo, for example, often introduce dialogue that intentionally highlights family structure, turn taking, respect, boundaries, care, and/or safety messaging or themes. Jewish parents who fit the profile of Facilitators, as another example, sometimes discuss totally different themes, perhaps situating zoo animals not as sentient beings worthy of care, but as kosher and non-kosher, or focusing on an animal's hoof to guide dialogue that will foster understandings of animal morphology that use a framework of religious doctrine. Facilitators who are parents may use a liminal approach for family instruction that assumes and/or reinforces sociocultural biases such as heteronormativity or life type hierarchies that intentionally and/or subconsciously frame conversations and concepts – sometimes in ways that do not match the scientific reality of the situation.[10]

---

[7] Tedlock and Mannheim 1995.   [8] Myers 2007.
[9] Allen 2002; Ash 2003; Leinhardt and Knutson 2004.   [10] Fraser and Sickler 2008b.

Previous chapters have pointed to the fact that human contact with animals creates countless points of entry to reflect on, discuss, and debate the common and distinctive needs and behavior patterns of living beings, as well as the responsibilities, outcomes, and opportunities that arise when humans impact spaces and systems. Visitors converse among themselves, and public audiences also authorize zoo staff to help them understand animals and related topics as wide ranging as water quality and how construction decisions impact animal life. WZAM2 data suggest that public audiences overall value opportunities to engage with zoo and aquarium staff and consider these professionals to be trustworthy conservation communicators.[11]

As such, zoos can leverage this opportunity to develop sites and staff training to facilitate productive conservation dialogue. These dialogues need not be instructive but can instead focus on the appropriate conditions for and moral dimensions of both conservation and animal care. We see these as connected themes that are (simultaneously) personal, communal, conceptual, *and* highly relevant. Beyond pointing public audiences toward information, resources, and complementary opportunities for learning, action, and engagement that might be of interest, zoos and other cultural institutions can help public audiences identify their questions and highlight values and strategies that support shared visions for the future.

Based on his experience as an advocate for open public conversations, museum and library scholar David Carr has noted the grave importance of intergenerational spaces, such as zoos, in which "discourse, caring, deliberation, reflection, and the constructive acts democracy makes possible" can be nurtured and normalized.[12] Carr has also found that quiet moments, confusion, and wordless experiences often indicate thinking and sometimes ground experiences that change a person's ways of understanding or navigating complex matters and uneasy tensions – which suggests zoos should thus anticipate and support these non-dialogic emergent outcomes, as well.

Ethics and social movements take form and gain power through language, which allows people to "objectify, label, and evaluate behavior," as

---

[11] Ibid. WZAM data suggest that the perceived value of zoo and aquarium staff as resources diminished for conservation topics less explicitly connected to animal well-being, though many respondents did value aquariums as a resource for broad (nonanimal) environmental topics such as pollution and water/energy conservation. Interestingly, educators valued both zoos and aquariums for providing information about a broad array of environmental topics slightly more strongly than the public overall but showed only moderate support for prioritizing the alignment of zoo programming with state education standards.

[12] Carr 2011, ix.

convergences of language and action support the development of shared moral concerns and norms.[13] While mechanisms of control, management, and collective action may well be appropriate and even necessary tools to pursue a conservation agenda, the conservation narratives we envision consistently situate individuals and community groups as important, relationally connected actors in conservation stories and outcomes. Indeed, we note that because nature and human–nature relationships are neither fixed nor uniform, dialogue facilitated by zoo staff should be flexibly initiated and adapted to fit the setting, participants, and participants' intentions, using the context as scaffold for diverse learners with varied expertise to progress toward new levels of understanding, commitment, and conflict resolution.[14]

The Question House Model

New York City zoos have long been notably intentional about finding multiple paths to use their civic role as trusted authorities to advance dialogue about (and acceptance of) new concepts regarding the rights of nature and the representation of responsibility to nature. One such early experiment featured specific focus on dialogue as a pathway for spreading important ideas with moral and behavioral implications.

During his tenure as Director of the New York Zoological Society, Fairfield Osborn recognized and honored the emergent learning that occurs within and among groups in the zoo, establishing Question House as a novel strategy to support young learners and their caregivers through visitor-initiated dialogue. Described in Annette Berkovits' recent memoir, *Confessions of an Accidental Zoo Curator*, Question House was a popular program that tasked committees of women volunteers throughout the 1950s with answering children's questions in an area of the Bronx Zoo designated for this purpose. These women were embraced as community resources engaging in undertakings "appropriate" for women in the United States during the post–Second World War period – but were not deemed sufficiently valuable to merit compensation or training. By the late 1960s, though, the Bronx Zoo had changed with the times, becoming one of the

---

[13] Myers and Saunders 2002.

[14] Carr 2011 specifies fifteen as the largest number of participants an "open conversation" group can accommodate before members of the group are likely to feel frustrated by challenges of speaking, listening, and responding as they might like. Such intimate experiences will not be possible for, or desired by, every zoo visitor, but could be tested and iteratively improved as a mission-supporting feature of zoo programming in many zoos around the world.

first zoos to pay staff (albeit poorly) to pursue institutional mission by educating visitors. Berkovits, who researched and led that visionary education program, maintains that *methods*, not content, most impact learners' attitudes toward animals and their habitats.[15] It is this data-supported understanding that now grounds ongoing efforts in and beyond the Bronx Zoo to effectively deliver fun, interactive, discovery-based conservation programming that allows for emergent dialogue, preferably with a member of staff on hand to facilitate and deepen visitors' discursive and reflective engagement.

Metaphor

It has become well established that emotional processing is a fundamental aspect of knowledge construction because emotions are a central means by which humans (and numerous nonhuman animals) interpret situations and share signals in social contexts.[16] Metaphor, the use of figurative language and imagery to transition feelings and emerging understandings from one distinct subject or focus to another, can be a tangible reflection of emotion–cognition as integrated, affective aspects of the learning process.[17] Put more simply, metaphor is the profound and frequent feature of human meaning-making that occurs through the comparison of one thing with another and complements logic on the continuum of language use, imagination, and rationality.

Commonly employed as key elements of discourse, metaphors are rhetorical devices used to convince an audience by casting an idea or situation in specific ways. Artistic and metaphorical ways of making meaning about animals, and affirming human–animal connections, vary, evolve, and are widespread in all cultures.[18] The common use of zoo-specific rhetorical devices speaks to the ease with which the human mind makes connections to zoo-related tropes of order and righteousness, and

---

[15] Berkovitz 2017.

[16] See, for example, Matthews 2004. Schutz et al. 2006 critique scholars' inadequate acknowledgment that emotions arise and play out within specific sociohistorical contexts of educational practice.

[17] Scholars recognize the attribution of human traits, emotions, or intentions to nonhuman entities – anthropomorphism – as one prolific cultural practice rooted and reflected in humans' systematic use of animals as metaphor; see Herzog 2010.

[18] While art was not the focus of this chapter, learnings from poetry installations in zoos in the United States can also inform agendas for the creative and intentional use of visual art as another tool that may shift perspectives or motivate zoogoers to further exploration or active adoption of a conservation ethic.

how frequently individuals and groups rely on, reinterpret, and contest metaphors human cultures have relied on for centuries. Furthermore, analysis of representations of zoos in the media and in popular literature can complement zoo visitor studies and discourse analysis as another way of exploring understandings of zoos' mission and institutional value.

In his classic dystopian short story, *Welcome to the Monkey House*, Kurt Vonnegut makes memorable use of zoo metaphor to explore and critique various intersections of human behavior, moralism, science, death, human–animal relationships, and concepts of free will and freedom. A common design trend when the book was first published in 1968, "monkey house" zoo exhibitions are the cultural phenomenon Vonnegut uses as a starting point for suggesting that the denial of cross-species similarities and connectedness is farce. He also leverages the zoo-as-chaos metaphor to highlight the arbitrariness, questionable legitimacy, and unintended outcomes of top-down imposed order and one-size-fits-all values.

It is, of course, the memorable embodiment of childlike innocence and trope of curiosity that husband-and-wife storytellers Margaret and H. A. Rey so effectively developed to position Curious George as a culturally iconic STEM ambassador. This primate protagonist always finds his own ways to science knowledge and moral reflection, often through experiential and coincidental pathways that parallel some specific science concept or disciplinary lesson he has been formally encouraged to learn, and often through direct or indirect intersections with the human and physical resources of the local zoo. By carefully situating Curious George as a metaphor for the "naturally" curious child in all of us, the authors seek to motivate inquiry, hands-on learning, and critical thinking.[19] Young children's production and grasp of metaphor actually tends to be superior to that of older children and adults, a recent finding that upends common assumptions about cognitive competency developments over time, and points to the importance of being intentional about supporting young learners' metaphoric strengths.[20]

Analyzing the multifaceted deployment of zoo metaphor by the media during the early stages of the US–Iraq War in 2003, anthropologist Kathryn Denning has shown how "the zoo" was presented metaphorically

---

[19] Schwartz-DuPre and Parmett 2017 offer a very interesting critical assessment of the *Curious George* children's book series as a conceptualization of STEM education as a crucial and ideologically neutral imperative in an era of ostensibly postcolonial global competitiveness that continues to be heavily impacted by the legacies and continuations of colonialism and colonial imagery.

[20] Egan 1997.

as representative of the invaded nation, degradation, and conquest, but also as symbolic of normality and freedom. Rhetorical uses of the zoo and zoo animals as symbols of other structures and outcomes thus followed either strongly positive or strongly negative (but always politically charged) patterns connected to simultaneous associations of the zoo as a context of harm, subjugation, and vulnerability, but also a context of crucial caregiving.[21]

Zoos and aquariums serve a clear metaphoric function as landmarks of animal care and civic pride. Maintained through the notable and ongoing responsible care for the well-being of captive animals and the provision of valuable services to nearby communities, credibility must be carefully maintained, lest the zoo-as-metaphor trope become connected with a specific zoo or the community or culture tasked with maintaining its credibility. The poor treatment or death of zoo animals are, in fact, frequently used by journalists and media outlets as metaphoric tools to illustrate urban depravity, inhumanity, moral superiority, or the decline of civility – suggestions they know tend to generate emotions such as concern, fear, distaste, or pride.[22] Metaphors of animal care and zoo settings are particularly common features of wartime narratives, consistently used to illustrate concepts and exemplars of morality and/or immorality. Pop culture examples of this pattern have recently included the best-selling novels *The Zookeeper's Wife* and *Babylon's Ark*.

Leveraging Discursive Mechanisms for Learning in the Zoo

WZAM2 data show that most zoo visitors believe a visit should include reflection on animal life as well as human responsibilities.[23] The role of critical dialogue as a tool for negotiating meaning and advancing conservation understandings and action thus becomes a critical consideration for those tasked with providing education programming in the zoo.

Exploring the role of dialogue at the intersections of thinking, persuasion, and attitude change, researcher Edith MacDonald recently collaborated with colleagues to use the elaboration likelihood model (ELM) to measure conservation messaging impacts at the zoo. The ELM is a theory of persuasive communication that suggests the value of inviting the listener to elaborate on a core messaging concept by scrutinizing what it might mean in context, in dialogue with a message communicator. The research team trained zoo staff to use this strategy of persuasion, recommend

---

[21] Denning 2008.   [22] See, for example, Dahlburg 1996.   [23] Fraser and Sickler 2008b.

specific behavioral choices, and invite dialogue. Consistent with other tests of the ELM model, survey data showed that visitors leaving the zoo were more likely to recall the behavioral message – and to understand it in context – than visitors who did not have access to this type of dialogue experience. Importantly, visitors found that knowing specific conservation behaviors they could pursue after leaving the zoo made their overall experience more satisfying.[24] MacDonald found the same high levels of satisfaction when pledge cards were used as a memory trigger, a strategy that also resulted in visitors' long-term retention of focal behavior messaging,[25] and Judy Mann-Lang and colleagues replicated these results the following year in a comparative study of dolphin programs.[26] Contrary to lore about visitors wanting entertainment, not "education," these results demonstrate that learning about conservation choices is a zoo experience visitors find satisfying, and that staff-initiated dialogue is an excellent pathway for achieving desired outcomes.

In addition to these encouraging findings, and the success of the already-mentioned open dialogue Question House model, work in the field suggests that metaphor can also be intentionally and effectively leveraged as a pathway for connection making and learning within zoo settings. Social scientists in the United States have shown that on-site poetry installations can facilitate affective learning and collective meaning-making that leads to new paths of understanding, qualitatively deepens visitors' experiences, and promotes conservation thinking. Not unlike the zoo, poetry makes heavy use of metaphor and can be produced and used as a tool and feedback loop for shaping and framing the self, and for making thinking and conscious and subconscious intentions more visible.[27]

One qualitative study of a curated poetry installation throughout New York City's Central Park Zoo assessed how site-specific metaphors influenced how 101 visitor groups (185 visitors in total) described wildlife conservation issues and characterized themselves in relation to nature using an open-ended group exit interview process that allowed for spontaneous discussion. The research findings demonstrated that the establishment of

---

[24] MacDonald et al. 2016.   [25] MacDonald 2015.   [26] Mann-Lang et al. 2016.
[27] Randy Malamud 1998, 2003 has explored numerous tropes human culture offers for the consideration of animals, including beliefs concerning animal souls and related aesthetic ideals for transposing animals into art and applying these ideals in animal poetry. The Mesoamerican spiritual system, Malamud claims, offers a standard that can be used as an appropriate point of reference to evaluate cultural inscriptions of animals, such as animal poetry, that have emerged in contemporary industrial–global societies, which tend to be less connected to ideas celebrating the sanctity and parity of nonhuman animals.

shared understandings about relationships between humans and human societies with the world of wild nature – and finding metaphors to articulate related concepts, perceived roles, and learning experiences – are significant aspects of the social experience of zoogoing. It is important to note, though, that project researchers also polled thirty zoo designers across the United States to determine their familiarity with the use and value of poetry as an integrated component of zoo architecture. Of the zoo designers sampled, twenty-eight had never used, or considered the use of, poetry in an exhibit.[28]

A follow-up project was initiated a few years later to ascertain whether the success of the Central Park Zoo poetry installation project could be replicated in other urban zoos in the United States and potentially strengthened through collaborative zoo–library programming designed to help public audiences further explore the intersections and parallels of literature and science. Findings again suggested that poetry complements the language of science in zoo settings by enabling shifts in perception that facilitate a sense of discovery and the alignment of feeling and learning.[29]

As they reflect on new data and underexplored pathways that might advance mission goals, zoos might look to satirist Kurt Vonnegut's *Welcome to the Monkey House* as a relevant point of reference for rethinking best practice around communication goals and structures. We offer this suggestion because we believe zoos should be seeking to provide public audiences with thought-provoking (but unresolved) conclusions to the very complex tapestry of emotions, thoughts, narratives, and values that emerge when zoogoers encounter animals, sociocultural frameworks, and other people (by intention and by accident) at the zoo, as Vonnegut was able to do through the medium of this clever short story.

---

[28] Fraser et al. 2008.
[29] See Preston 2013 for details about the goals, methodology, findings, and implications of this endeavor to explore whether the initial Language of Conservation poetry installation project in Central Park Zoo could be replicated in other US cities and library–zoo partnerships.

Bonding

A Sociobiological Human Need with Important
Zoo Mission Implications

A conservation biologist whom John Fraser was able to interview as part of a study about early childhood memories of visiting zoos described the zoo as an important part of a childhood condition that required regular appointments with the doctor. Following these appointments, which included injections and other discomforting procedures, his mother would take him for ice cream at the local zoo. These were remembered as particularly important moments because they involved impactful learnings about his mother's values and beliefs about how and why it was their duty to care for wildlife. Reminiscing about childhood camping trips with his father as another important learning opportunity, he mentioned that his mother had always joined them for daytime camping experiences but would take her leave as evening fell to spend the night at a nearby hotel. He learned later in life that his mother greatly valued that family time in nature but did not wish to camp because of complex memories of her own childhood, part of which had included poverty and living in migrant worker tents with her parents and siblings.

As this well-known conservationist relayed stories of his childhood and reflected on various moments and patterns that had been impactful, he described zoo experiences with his mother as deep discussions about shared interests and life experiences. Affirming that animals frequently sparked their shared inquiry and dialogue, he mused that he and his mother had often been peers learning side by side and declared that these experiences grounded a unique bond between them that has continued to inspire him throughout his life. After his mother's funeral, he returned to the zoo and sat on a familiar bench to reflect on the special relationship they had.

All humans belong to social groups, social relationships, and cultures that help them feel secure, recognized, and understood. About fifteen years ago, scholar Maximillian Holland's ground-breaking interdisciplinary theoretical analysis clarified the utility and limits of applying insights from biology to understand human social behavior at the individual and primary social group levels. Holland established that while human social bonds and kinship structures are neither bound nor dictated by genetic relationships, sociocultural and contextual cues do clearly mediate expressions of social behavior.[1] While the dynamic, interpersonal, emotional process of developing relational, affiliative attachments (bonds) can develop in numerous settings, cultural institutions are recognized as particularly suitable contexts for human social strivings and flow experiences because these settings are intentionally designed to provoke people to clarify and interpret concepts, and weave new information into existing ideas and frameworks for understanding themselves and their shared world.[2]

Indeed, a unique aspect of most human bonding experiences at the zoo appears to be the opportunity to look for, observe, feel part of, reflect on, and discuss *zoo animals'* pursuits and expressions of love, connection, and affinity. As they process such phenomena, zoogoers frequently engage in dialogue with other visitors as co-learners, and this process of mutual learning facilitates trust, social bonding, and cohesiveness by promoting shared understandings or perceptions of identity, temporal order, and causal connections. We seek here to introduce some of this scholarship, teasing apart the unique conditions that appear to strengthen social bonds within zoogoing groups[3] and highlighting pathways that appear to foster shared perceptions about valuing and caring for animals and nature.

Human Bonding Pursuits and Outcomes at the Zoo

To better understand social capital, a multidimensional concept typically understood to describe the connections, shared values, and shared understandings that enable trust and collaboration,[4] several researchers have explored zoos as public venues that offer diverse audiences accessible content associated with topics of shared concern. Recent data suggest that politicians and faith leaders alike consistently recognize zoos as valuable, as

---

[1] Holland 2012.

[2] Csikszentmihalyi 1990. For a parallel assessment of why social bonding in public, private, and nonprofit institutional contexts must be (re)examined as an important aspect of processes to strengthen democracy, encourage civic engagement, and build better communities, see Gates 2003.

[3] While parallel patterns of social bridging, the process whereby members of different visitor groups connect and discover areas and possibilities of trust and reciprocity, is not the focus of this chapter, social bridging is also common in zoos; WZAM data suggest that while visitors do not explicitly come to the zoo to interact with strangers, they do tend to feel that zoos are one of the most comfortable venues for engaging in bridging experiences (Fraser and Sickler 2008b).

[4] Robison et al. 2002.

public settings that draw diverse groups together and are widely perceived as having legitimate authority on matters of community, culture, and values.[5] Even national-level assessment has identified zoos as venues that facilitate social capital development by allowing visitors to build relationships and negotiate meaning in ways that often lead to the construction of group-level social knowledge construction and shared values – processes important to the widespread adoption of a conservation ethic and pursuit of conservation outcomes.[6]

Building on the pioneering work of researchers such as Susan Dale Tunicliffe, the WZAM research agenda included exploration of how zoo-visiting family, school, and affinity groups, as well as another notable "user group," zoo volunteers, appear to be impacted by social bonds forged through zoo experiences. The researchers found that trust-building consistently emerged as a highly valued outcome of social bonding during group zoo visits, though the primary motivations that provoked and informed various groups' zoo visits unsurprisingly differed by group type. School group leaders, for example, were explicitly education focused, whereas family group leaders tended to be value focused, and while family model concepts were the overwhelming focus of caregivers with family members in their visiting group, community or social model concepts were often the main focus of school and affinity group zoo visits. Zoo volunteers, interestingly, had different perspectives on the role of social bonding in zoo settings.[7] To begin to situate these nuanced patterns and tease out potential conservation mission implications, we outline research frameworks, data, and observations that point to different conceptualizations and outcomes of social bonding that impact various groups of zoogoers.

Family Bonding

Observational study of parenting behaviors in zoos has pointed to an underlying pattern of purpose related to social bonding that characterizes most caregivers during family visits: the reinforcement of social boundaries, family membership, and why such dynamics are "natural" and

---

[5] Fraser and Sickler 2008b.

[6] In 2000, researchers Jenny Onyx and Paul Bullen developed a set of testable constructs related to trust, reciprocity, socially proactive behaviors, and community network participation. National-level assessments of social capital through surveys of cultural participation in museum programs across Australia, based on Onyx and Bullen's framework, specifically highlight zoos as venues that build social capital (Australian Bureau of Statistics 2004).

[7] Fraser and Sickler 2008b.

important. Fraser and other researchers have noted that caregivers in the zoo can frequently be seen and heard helping children understand choice, autonomy, social aspirations, build connections with others, learn responsibility, and establish opportunities for shared enjoyment.[8] Such patterns suggests that zoo visits have an important role in the development of structural and systemic relationships. Research on the use of socialization at zoos to build positive regard for community[9] and provide valuable out-of-home experiences that can establish children as part of a community[10] has shown that zoos can be exemplars of public venues that offer diverse audiences accessible content that relates to topics of shared concern.

WZAM social capital studies data confirm that family visits typically feature, as an a priori goal, the intention to solidify the family bond. During a qualitative interview, for example, one mother explained, "I don't really like zoos, but it's worth it for the time with my son." Common rituals among family groups include caregivers using animals as models to talk about and provide instruction regarding "appropriate" bonding, affiliation, and touching, often in order to establish specific understandings about order, hierarchy, continuity and commonality, defending oneself and others, and roles and responsibilities. That is, while parents visiting the zoo with small children tend to spend a lot of energy directing their children's gaze toward the displays and the animals therein, socializing them to pay attention to what the zoo seeks to prioritize and communicate, parents also spend a lot of time directing their children to take turns, say please, notice how smaller animals appear to be showing respect to larger animals (perceived to be parents or elders), notice how an animal's behavior is like or unlike human behavior, etc.

When WZAM researchers spoke with family groups about their experiences with a specific exhibit, it was extremely clear that families had a wide range of different conversations and takeaways that were meaningful to them and revealed how they were using and valuing the zoo. A few parents described having experienced abandonment in their own lives, and how that negative emotional experience had led them to intentionally emphasize "family" groups of animals to exemplify the importance of strong family bonds, describing to their children how love and mutually supportive relationships are found throughout the animal world and a best

---

[8] See, for example, sociologist Marjorie DeVault's 2000 research on public behavior among families visiting zoos.
[9] Longhurst et al. 2004.   [10] Chase 1993; Carter 2003.

practice. Another set of parents focused on the value of the zoo as a place to introduce nature to their urban offspring, who have little connection to the natural world.[11]

In 2009 John Fraser used a constructivist grounded theory approach to further explore parenting perspectives on the value of zoo visits undertaken by eight families who frequently visit zoos with their children and enrolled them in a zoo-based charter school that incorporated zoo programming into the curriculum and weekly student visits. Parents in the study sample believed the animal behaviors they saw with their children during zoo visits served as metaphors for peacefulness and accepted codes of conduct for affection and support. While some parents reported actively using and valuing the zoo as a tool to help their children reason about their role in human social situations, all parents in the sample felt that discussing zoo animals as models of altruism facilitated the purposive transfer of environmental values related to the appreciation of nature. The advancement of childhood moral development was frequently tied to bonding, as well as to connectedness to animals (the topic of Chapter 8). Altruism, described by these parents as a learned skill value set that includes compassion, caring for others, and empathy, was described as growing from "natural" compassion for individual animals that can be refocused and extended to include categories of animals as well as unknown people. Interestingly, while zoos are generally perceived to promote nature conservation and the parents in this study agreed with that characterization, they reported choosing the zoo as a frequent destination because the zoo allowed them to use animal behavior as an instructive tool to promote their own environmental values.[12]

Fraser has spoken with social workers about their use of these zoo settings in various locales in the United States as an ideal context for helping families work on reintegration strategies after children have been removed from the home. Given the zoo's status as a public setting where social behavior norms are a common topic of focus and conversation, social workers describe themselves as mediators who use various themes that seem appropriate in the zoo to frame the on-site experiences of family members with diverse psychological and social needs. We similarly know from WZAM parent interviews that caregivers whose families are working

---

[11] Fraser and Sickler 2009.

[12] Fraser 2009a. Grounded theory approach is a non-hypothesis-driven qualitative method integrating discussion and observation that is recognized as useful for uncovering new theory that can be subsequently appraised for explanatory coherence. See, for example, Strauss and Corbin 1988 and Charmaz 2000.

through abandonment issues come to the zoo hoping to observe and discuss with their children relational love and care between the captive animals as a theme central to the family bonding experiences they hope the visit will facilitate. Parents concerned that their children are unconnected to specific norms they themselves grew up with, on the other hand, visit the zoo with the intention of (re)connecting their children with wildlife and harmonious natural systems.[13] Different families visiting the same exhibit will inevitably thus create (and typically co-create) different meanings and values informed by specific aspects of their experiences that seem most closely aligned with their own immediate or longer-term priorities and goals.

School Group Bonding

For many decades, zoos have sought to situate their spaces as public, local sites for embodied inquiry-based learning and relevant personal connections. Community-as-classroom frameworks offered through the Calgary Zoo and Vancouver Aquarium, for example, seek to create place-based learning opportunities and develop ecological literacy and a conservation ethic among participating teachers and students by developing programming for extended, co-created, hands-on experiential learning. Such programming is often pedagogically linked to psychologists' understandings about the development of human cognitive process over time and the importance of engaging the body, emotion, and imagination in the local natural and cultural context in which students live and learn.

Based on his ongoing exploration of how educators and institutions of learning might optimize human learning capacities after early childhood, education analyst Kieran Egan has identified stimulation of the imagination as the best pathway for preserving intellects sufficiently flexible to effectively address the complex demands, opportunities, and challenges of the present and the future. Egan describes the development of cognitive process as an individual's mental framework shift from general understandings of hierarchical relationships that support the adoption and internalization of social group patterns toward (new) brain functions that support understandings of the abstract and relational domains. As examples of this cognitive spectrum, Egan outlines a series of literacies as intellectual tools

---

[13] Fraser and Sickler 2008b.

that build on foundational mythic understandings of the world that typically emerge in early childhood and support imaginative flexibility later in life. The romantic literacy that typically emerges as a marker of deepening social intelligence during the teenage years, for example, is characterized by more nuanced understandings of reciprocal relationships and is often facilitated by conditions that are exotic or awe-inspiring, according to Egan. Philosophical understandings, which can emerge as a complementary phase of cognitive development, arise through the pattern of searching for patterns and constructing and revising theories to make sense of general truths. Well-developed ironic understandings, which are characterized by intellectual and ethical reflexiveness and represent a deeper level of cognition, require that sympathy and sensitivity be extended not only to oneself, but also to seemingly dissimilar other beings.[14]

Zoo experiences inevitably disrupt, for most zoogoers, pedagogies, sociocultural narratives, and disciplinary traditions that situate science and rationality as a realm of disembodied cognitive intelligence entirely distinct from affective, sensory, humorous, and imaginative experiences. John Fraser has discussed with many educators over the years how much they value zoos as sites where new stimuli are likely to provoke shared learning and bonding experiences that push students to engage their awakening romantic, philosophic, ironic, or other developmentally relevant capacities as they process zoo experiences together. To support and provoke multiple understandings in a zoo setting a teacher or zoo staff member could, for example, guide a group of adolescent students to observe and reflect on the incredible mental and social capacities of captive primates, encouraging interdisciplinary and naturally emergent dialogue and laughter about a range of related topics such as behavior patterns in various types of animals, including humans; how humans better understand ourselves and stories when we learn about and from other animals; the students' emotional responses to various exhibits; and/or the ethical implications of human behaviors driving biodiversity loss and climate change. Whatever form it takes, a holistically productive learning experience might be understood to have transpired if a school group leader observes widespread affective engagement and imaginative involvement; social bonding; and use of questions, comments, or jokes that suggest students are attempting to apply cognitive tools to make sense of experiences in the zoo during or after the visit.

[14] Egan 1997.

Affinity Group Bonding

While theorists assert that shared affective energy can contribute to social cohesion, as discussed in Chapter 5, sociologists have also analyzed group pleasure through emphasis on the impacts of shared affective energy on the establishment or reinforcement of a common focus.[15] Fun zoo experiences, including dialogue, related or tangential to a shared affinity can solidify group identity and can also reinforce or deepen group boundaries and norms – and such processes are interwoven with meaning-making and learning for affinity group visitors. Recognizing this, many leading zoos actively collaborate with interest groups to reach specific public audiences.

The Oregon Zoo, for example, addresses lead as a hazard to humans and wildlife alike through education and outreach programming designed to engage hunters, whom the zoo situates as being among the most supportive voices for wildlife and habitat conservation. While hunters might (also) bond over complementary or shared domination narratives, care, dependence, and connectedness tend to be the key foci of bonding conversations initiated in zoo settings, and the Oregon Zoo seeks with this programming to deepen the conservation identity and behavioral norms of this affinity group by inspiring them to choose non-lead ammunition, which has been shown to reduce the threat of toxic lead exposure to wildlife.

Bonding among Staff and Volunteers

Research and observation suggest that different types of bonds tend to develop between different worker cohorts in zoo settings. There are, of course, many types of work happening at the zoo. While some staff provide visible and straightforward on-site services, others, such as senior managers, exhibit designers, landscape architects, and media relations staff, have visionary operational roles that often unfold behind the scenes. Here, however, we will focus on the bonding of the paid and unpaid workers who engage directly with animal care and/or conservation education, work that can include delivering structured and impromptu educational content, developing curricular and program materials, or engaging learners through animal encounters or other animal-focused learning or activity opportunities at the zoo or in local communities.

Recent survey and interview data show that zoo staff and volunteers tasked with visitor communication regarding animal care and

---

[15] See, for example, Mead 1934.

environmental advocacy tend to feel that their in-group status as front-line conservationists separates them in fundamental ways from society at large and even from other worker cohorts at the zoo. Those tasked with visitor communication regarding animal care and environmental advocacy consistently report feeling that their in-group status as front-line conservationists separates them in fundamental ways from society at large and from non-conservation-committed individuals and groups they interact with at work and in their daily lives.[16] Several sociologists have noted that many zookeepers in many locales act goofy and sing silly ditties to express their appreciation for and tease each other, often using inside jokes to mark community.[17] While silly and teasing interactions may also occur between – and be beneficial to – groundskeepers or ticket sales or food service staff providing important services at the zoo, there is reason to believe that animal care and conservation educator staff and volunteers build particularly deep and impactful cohort bonds that validate the meaningfulness of their group. Describing zookeeper bonding, sociologist David Grazian notes, "Birds of a feather, keepers forge loving relationships with zoo animals and each other, collectively animating their work with shared meaning and moral purpose."[18]

While it is satisfying to share meaning and moral purpose, it is also the case that strong awareness of the causes and consequences of environmental degradation coupled with the perception of political and social indifference within one's immediate social network can become an ongoing trigger for psychological stress and damage the emotional well-being of animal care and conservation educators. The emotional toll of environmental education work can be intense, and indicators of depression, anxiety, and enervation are significantly elevated among conservation educators. A growing number of psychologists have come to the conclusion that active conservationists consciously and/or subconsciously live with a high degree of negative emotional experience because they are acutely aware that mainstream policies and practices are incredibly disconnected from the conservation ethic and outcomes they seek to advance.[19]

Data analysis from interviews of climate science researchers and staff responsible for interpreting the causes and implications of climate change

[16] Fraser and Sickler 2009; Fraser et al. 2013.
[17] See Fine and Corte 2017 for an assessment of recent research around how the sociology of fun appears to intersect with shared narratives and collaborative commitment in zoos and other settings.
[18] Grazian 2015, 14. Grazian is also an excellent starting point for scholars and laypeople interested in the related, ongoing feminist story of women's strikingly low status and low pay work in zoos.
[19] See, for example, Lipsky 2009 and Fraser and Brandt 2013.

in informal learning settings such as aquariums, science centers, and conservation sites points to specific patterns that appear to be associated with distress, feelings of hopelessness, and heightened sadness: explicit acknowledgment of the reality and implications of climate change, lack of social support, and persistent negative messaging. While engaging in conservation education work has been associated with feelings of grief, strong desires to avoid confrontation, and stress from constant reminders that biodiversity and ecological sustainability are being constantly and rapidly eradicated throughout the world, most of the informal educators in the studies highlighted here reported that peers who did the same type of work were a crucial source of psychological support and stress mitigation. The combination of perceived urgency of conservation education work and the constancy of perceived threats to success appear to lead to tight bonds within the community of zoo staff passionate about conservation.[20]

WZAM findings demonstrate that social bonding within the volunteer corps tends to serve as a support system and reinforcement mechanism, and to facilitate social action. While camaraderie and collective belief in a shared emotional register can increase the fun and solidarity of any group, the peer support system many zoo volunteers describe has other tangible benefits, as well: Bonds among unpaid staff at the zoo form the basis of a self-reinforcing network to promote shared values-based action, according to interview respondents. Zoo workers who perceive strong bonds with their colleagues more frequently report that they feel capable about making a difference with respect to conservation outcomes than zoo workers who do not feel they are part of a cohesive group. Those in the former situation often describe taking up the work for the animals but staying with it for the people. Beyond using various bonding strategies and drawing on established bonds to offer mutual support when animal care is disgusting or conservation communication work is frustrating or depressing, zoo volunteers and staff consistently describe how they built up a social cohort over time that now makes them more comfortable committing to the more active pursuit of a more outward-facing conservation agenda.[21]

Furthermore, conservationists who perceive that their family and/or coworkers share their environmental values report less fear and panic than those whose immediate personal relationships do not ameliorate the weight of their environmental knowledge and beliefs. Given these emerging understandings about various tools and conditions that might improve

---

[20] Fraser et al. 2013.   [21] Fraser and Sickler 2009.

the well-being of conservationists working in zoos and other informal environmental education settings, scholars have begun to focus on psychological processes and aspects of professional training and development that might help environmental educators build resilience and maintain hope, a topic to which we will return in Chapter 11.

Different Folks, Different Strokes

Though WZAM data confirm that conservation-committed zoo staff and volunteers sometimes feel frustrated and isolated when they see other staff and visitor groups engaging in zoo experiences that appear to them to be off-topic, such as social bonding or pleasure seeking, the data presented in this chapter support our overall conclusion that a very wide and often layered range of zoo experiences can be consistent with and even advance a conservation agenda. While critics often point to the lack of tangible science content learning that can be directly connected to a zoo visit, researchers have established that social interactions around animal exhibits often strengthen social bonds that people learn from and through, and such experiences can impact the establishment of shared conservation values because in-group bonding and general identity explorations are typically precursory to the adoption of a conservation mindset.[22]

While this chapter has focused on the nature and implications of in-group bonding priorities and outcomes often identified as particularly important dimensions of zoogoing, other psychological functions fulfilled by zoos may also contribute to high attendance – irrespective of, but sometimes convergent with, the formal mission goals of zoo operators. In addition to using zoo experiences to illustrate generalized concepts about human socialization, scholars have proposed that zoos may fulfill an innate need for affiliation with living nature and support identity development. In subsequent chapters, we assess these claims and establish what has been shown about the psychological services and benefits zoos may provide, beginning, in **Chapter 8**, with the unique affordances of zoo animals as beings many zoogoers consider themselves to be connected to and representative of larger groupings of animals potentially worthy of active care.

---

[22] Falk 2005.

CHAPTER 8

Connectedness
Animals, Continuity, and Belonging

Maria, her husband, and their teen daughter and tween sons were visiting their local zoo in New York City. They lived within their means in low-income public housing a few blocks from where Maria and her husband had themselves been raised. Their urban neighborhood was one of the few in New York City that continued to struggle with gang violence and an active, illicit drug trade. Maria's father had left the neighborhood to live in a farming community four hours north of the city.

Maria felt strongly that her parents' strong moral values and religious education were key factors that had helped her grow up with and maintain a deeply ingrained moral code. She described their present-day situation and environment as hostile toward children and explained that these conditions motivated her to devote a great deal of her time to volunteering at her youngest son's school. She felt that the conflicts her family and community were forced to navigate every day situated them as abandoned and "othered" in a way that was out of balance with the natural order of the world. Maria offered a notably philosophical view on the environmental values she sought to instill in her own children through zoo visits.

> *When we come to the gorillas, they see that part of the mother, the nurturing part of the mother. As a matter of fact, they do say how it's so compatible . . . the nurturing part . . . the way I am with them, the way they see the animals are so bonded. I mean, a life partner, like the penguins, you have that life partner. It's the same role that we're trying to show our children. An animal takes that life partner to be with them for the rest of their lives. They're trying to teach their kids, the same way we are. When it comes down to it, it's the family, it's the unit. It's very important to us.*

Maria made it clear that she used zoo exhibits to illustrate a desirable, idealized natural order. She saw her role as facilitating the moral development of her children by presenting concrete examples of how human relationships depend on the development of trust, reciprocity, and selflessness to protect group members from external threat. For Maria, the natural order of the world points to and is an example of the continuity of all forms of life, a concept and point of connection she found comforting and important to share.

Human Connections to Animals and Nature

Often tacitly experienced, connection and belonging are powerful dynamics of emotional affiliation that are notably difficult to articulate.[1] Humans often seek out experiences that establish or reinforce their feelings of belonging, in many cases by affirming their sense of biological continuity and their affinities for and affiliations with animals (both human *and* nonhuman).

While a compelling array of international and interdisciplinary scientific studies now suggest that nature may be a human necessity, rather than a luxury, given the range of physical and cognitive benefits that have been shown to accrue to humans who connect with nature,[2] there is no evidence that human connections to animals and nature are biologically determinant, despite significant scholarly focus on this topic for the past thirty-five years. There is, however, overwhelming evidence that experiences with living things, be they plants, natural areas, wildlife, or pets, have direct psychological value. Here, we thus outline some of the key philosophical and research underpinnings that inform current understandings in order to situate various implications with respect to the social value of zoos.

A Short Summary of the Biophilia Hypothesis

In 1984, entomologist E. O. Wilson proposed the idea of biophilia, the idea that humans have an innate emotional need for affiliation with other living organisms. He proposed that we are evolutionarily predisposed with a desire to affiliate with the natural world – citing high attendance at zoos as one possible illustration to support his thesis.

While Wilson's premise was a bold integrative idea, it now seems a logical extension or outgrowth of the British romantic tradition of the 1800s, characterized by the return-to-nature movement, which positioned

---

[1] Wood and Waite 2011.
[2] Florence Williams highlights a wide range of cutting-edge science research from around the world in her 2017 journalistic exploration into nature's restorative benefits, citing evidence that human connections to nature are essential to human cognition and spending time in natural environments tangibly impacts human well-being in numerous ways.

a nostalgic view of the countryside as a restorative response to the toxic threats of industrialization. In Wilson's context, a century later, the modern environmental movement had been gathering momentum across the United States for decades, in the face of widespread toxic damage being caused by global industrial pollution. James Lovelock and his colleague Lynn Margulis had recently proposed the Gaia hypothesis in the mid-1970s, suggesting that the world operates as a self-regulating system of feedback loops that sustain life as a single, interconnected organism.[3] This idea, which became a central pillar of thinking about the value and credibility of a whole-earth approach to understanding nature and the environment, percolated for a decade before emerging as a popular book and entering the conscience of the general public. An integrated explanation of human life and biosphere processes, the Gaia hypothesis informed a new paradigm of interconnectivity in which the continuity of life and affiliation are core aspects of an overarching feedback loop.

Taking this line of thinking in a new direction, Wilson's biophilia proposition was introduced as a polemic, and there has been some evidence supporting biophilia as a potential human trait, at a primal level. Evolutionary psychologists have shown, for example, that the human visual system features specific capacities to detect unexpected animals in environments with multiple animate and inanimate stimuli, irrespective of whether the animal is no threat (a bird) or a potential threat (a snake).[4] Developmental psychologists have also shown that – even when attractive toys are present, and even if the live animal options are potentially threatening (snakes and spiders) – infants, toddlers, and preschool-age children consistently choose to focus on and interact with the live animals more often than the toys, and also behave differently toward the animals, talking about the animals more and asking more questions about the animals than the toys.[5]

Wilson's premise suggests that zoos have been and will be enduringly popular because visitors instinctively value proximity to animals of all types. Over the years, though, critics have noted the lack of proof that decisions to spend time in zoos are driven to a large extent by visitors' conscious or subconscious built-in desire to affiliate with animals. Continuities between pet animals, domesticated animals, food animals, and wild animals also do not fit tidily into narratives about the "affiliative" domain of desire. Indeed, why humans seek affiliation with exotic captive animals when domesticated, local, and pet species offer similar

---

[3] Lovelock and Margulis 1974.   [4] Calvillo and Hawkins 2016.   [5] LoBue et al. 2013.

intersubjective opportunities without the risks and complexities of care management remains an open question.[6]

In a series of essays in the early 2000s, social psychologist Gareth Davey pointed to the possibility that simulated nature experiences assure people of the naturalness of the experience but do not contain the affordances necessary to provide the full feedback loop that would be required to meet the terms of a biophilic affiliative desire.[7] Two key questions for zoos today, then, seem particularly relevant. If the human desire to affiliate with animals is, indeed, a phenomenon (even if this desire proves not to be universal and/or innate, or to extend to all types of animal), how can this motivational pattern and desired outcome become more central to the mechanisms of mission delivery in zoo settings? Furthermore, do the highly constructed landscapes in zoo settings sometimes hinder emotional affiliation by limiting proximity and/or naturalness?

Following Wilson's initial biophilia proposition, several scholars considered whether zoos can rightly be considered culturally valuable for providing affiliation-with-nature services to their communities. Twenty years of original research that included zoo visitor studies led Stephen Kellert in 1996 to identify nine basic value orientations that people hold regarding wildlife and biodiversity, reflecting fundamental ways of attaching meaning to and deriving benefit from the natural world. Described as biologically based human tendencies moderated by culture, learning, and experience, Kellert's often-cited list of the broad value orientations that seem to consistently impact human perceptions regarding the importance of biological diversity include scientific, aesthetic, humanistic, moralistic, naturalistic, negativistic, dominionistic, symbolic, and utilitarian dimensions (see Table 8.1).

Claiming that our physical health, and capacities for affection, aversion, intellect, control, aesthetics, exploitation, spirituality, and communication are directly grounded in our connections to the natural world, Kellert has pointed out that a new ethic will need to become mainstream to resolve our current, interconnected environmental and social crises, motivated not

---

[6] As a starting point, psychologist Gareth Davey's 2005 literature review outlines various themes of critique that might inform development of an assessment framework for future biophilia research in zoos. Davey suggests, in particular, that researching visitor preference for more naturalistic exhibits and whether visitor experiences with more naturalistic exhibits directly benefit psychological or physiological health would be most be useful for demonstrating why people value zoo experiences despite vast differences in design pedagogy and degree of "naturalness" visitors might associate with different zoos.

[7] Davey 2005, 2006.

Table 8.1 *A typology of basic value orientations toward wildlife and biodiversity*

Value	Definition
Utilitarian	Practical and material exploitation of nature
Naturalistic	Direct experience and exploration of nature
Scientific	Systemic study of nature through the lenses of structure, function, and relationship
Aesthetic	Physical appeal and beauty of nature
Symbolic	Use of nature for language and thought
Humanistic	Strong emotional attachment and "love" for aspects of nature
Moralistic	Spiritual reverence and ethical concern for nature
Dominionistic	Human mastery, physical control, and dominance of nature
Negativistic	Fear, aversion, or alienation from nature

Source: Kellert 1996

by an abstract desire to "save" nature but by pursuit of rudimentary self-interest.[8] While Kellert specifically stated in his findings that most of the zoogoers he studied did not demonstrate enhanced appreciation for wild-life as a result of their one-day visits, he noted that more data would be needed to ascertain whether specific value orientations toward nature appear to lead to greater desire for zoogoing as an affiliative opportunity, concluding that zoos can potentially fulfill visitors' biophilia desires if exhibits are designed in ways that more clearly connect animal and human experiences.[9]

Zoogoers seek to connect to animals in numerous ways that suggest they value perceptions that they share experiences, thoughts, and feelings with captive animals.[10] People in the zoo wave at animals, try very hard to get the attention of and establish eye contact with captive animals, compare themselves (or people they know or humans overall) to animals, imitate animals, and speak on behalf of individual animals.[11] Anthropomorphism, the tendency to project human characteristics on nonhuman animals, is a

---

[8] Kellert 2012.   [9] Kellert 1997.
[10] John Berger 1980 offers a fascinating snapshot of humans' search for meaning within and behind the ways we look at animals.
[11] While the pursuit of these types of human–animal connection or acknowledgment could potentially deepen an individual's outlook and perspective of dominance (a point of concern central to some anti-zoo narratives), those who study empathy suggest that these connections tend to be about acknowledging mutual respect for another's being. This distinction in how connection is inferred is found in a variety of studies of zoo visitors and different types of zoo animals.

response to wildlife that has, in fact, been identified as a genetic tendency that appears to be universal.[12] However inaccurate, humans' belief that they are able to understand animal thoughts and experiences are an excellent foundation for building or affirming empathy and concern toward individual animals and perhaps animals and natural systems more broadly.

In 2005, conservation biologist Kristen Lukas and animal behavior specialist Stephen Ross built on Kellert's work by demonstrating that the majority of their zoo-visiting research participants reported affiliation with animals as an underlying reason to spend time in zoos, and expressed "naturalistic" values in their patterns of response to proximate experiences with captive African apes (which, in some cases, included limited recognition of personal kinship with these primates). While their findings suggest that many zoogoers may value zoo experiences for facilitating a widespread human desire to explore perceptions of self within a framework of biological continuity or biological kinship, the research design did not delve deeply into the conditions or implications of human–animal kinship exploration patterns in the zoo.

While there is still scant scientific evidence to support Wilson's claim that humans *need* to be intimately connected to natural environments, Peter H. Kahn, Jr.'s studies of various age cohorts in several countries have affirmed that universal patterns do appear to characterize humans' development of moral relationships with nature, and that children across diverse cultures engage in remarkably similar processes of environmental moral reasoning.

Investigating whether benefits that accrue when humans are in the presence of nature entities might allow biophilia to be studied through the lens of structural development, Kahn has proposed that benefits to humans who have proximate experiences with nature could be measured at an egoistic level and through the extension of their self-concept to include other natural entities such as animals. He suggests that biosphere-level benefits can be understood to occur only if an individual's physiological or psychological response(s) to experiences in/with nature led to tangible

---

[12] See de Waal 2019 for an overview of the Greek origins and historic uses of the term "anthropomorphism," a term now generally used to censure the attribution of human-like traits and experiences to other species, often to distinguish human cognition or recognize continuity in cross-species emotional responses. Based on his own experience as a committed and active conservationist, De Waal notes that this tendency is too often seen narrowly as problematic when it often helps people connect with animals in meaningful, even if inaccurately understood, ways.

accrued benefit to a natural system on which humans depend, and has also claimed that zoos teach children dominionistic values that are counterproductive to zoos' conservation mission.[13]

Some empirical evidence now supports Kahn's premise that biophilia is linked to biocentric educative benefits at the egoistic level. In 2007 and 2008, researchers used an implicit associations test to demonstrate that zoo experiences contribute to feeling connectedness with nature. In the initial study, data from repeat testing of visitors spending the day at the San Diego Zoo was compared to data from individuals who took daytrips to the beach or spent an afternoon on the golf course. The data showed that a day at the zoo increased visitors' tendencies to implicitly associate themselves with nature more than the other settings.[14] The follow-up study, which compared data from very different zoo types, including zoos with a natural garden design aesthetic and zoos featuring more classically built exhibits with glass dioramas of natural settings, resulted in similar outcomes, irrespective of the degree of naturalness represented.[15] While these findings demonstrate that arousal and association with nature are often enhanced by zoo experiences, neither study provides sufficient evidence to support the premise that biophilia is an innate human desire; shows that all humans desire such experiences; or sought to investigate whether biophilia is the predominant motivational driver of zoo attendance. It may be that, like the man profiled in Chapter 5 who habitually visited the Central Park Zoo during his lunch hour, zoogoers who leave their zoo visits with a higher personal association with nature than they arrived with may have experienced and appreciated the zoo as a way of "resetting their nature barometer" through conscious awareness of a set of cognitive values, rather than any biological human function.

Attitude and Worldview Analysis: Affiliation with Animals

The relationship of human societies to nonhuman animals has always been important to how nature is defined, valued, and used. While some moral principles may be universally recognized, most are notably fluid and socially negotiated – and will likely be expressed differently across different cultures. Social psychologists in the United States, for example, have suggested that public perceptions of wildlife are influenced by a combination of simultaneous and sometimes conflicting value orientations. Based on emergent data, researchers have proposed, for example, that, as

[13] Kahn 1999.   [14] Schultz and Tabanico 2007.   [15] Bruni et al. 2008.

Americans became more urban and more educated, value orientations toward wildlife and biodiversity have become comparatively less utilitarian and more mutualistic (mutualistic values being conceptualized as an integrative realm of concern that generally corresponds to Kellert's moralistic, naturalistic, and humanistic value orientations).[16]

When a team of researchers from Ohio State University replicated in 2014 a survey comprised of twenty-six single-item measures of attitudes toward animals that Stephen Kellert and Joyce Berry had used in a 1978 study of over 3,000 United States residents, they found that the 1,287 respondents in their own nationally representative sample reported remarkably similar attitudes toward domestic, companion, game, and predatory species representing a variety of taxa and varying degrees of known benefit or potential harm to humans. Finding that the greatest pattern of difference between the two studies highlighted significantly more positive attitudes toward historically stigmatized wildlife species such as bats, rats, sharks, vultures, wolves, and coyotes, the researchers concluded that this pattern was likely a reflection of shifts in public concern for animal welfare over the three and a half decades between the studies, noting that scientific study and dissemination of information about these species increased dramatically during that time period. Indeed, policy evidence demonstrating public concern for the individual well-being of wildlife began appearing in the United States in the 1990s and the researchers note that their study findings mirror an increase in positive attitudes toward sharks in the United Kingdom attributed to growing scientific and ecological interest among the public at large.[17]

Psychologists Alice Eagly and Shelly Chaiken define "attitude" as a "psychological tendency that is expressed by evaluating a particular entity with some degree of favor or disfavor."[18] Three components of attitude tend to be distinguished by psychologists: cognition (beliefs about the object in question), emotion (emotional response to the object), and conation (behavioral tendency toward the object).[19] Overall increases in attitudes that reflect concern for animals and inclination to explore nature, now that a greater proportion of society lives in urban settings (which offer fewer opportunities for direct engagement with animals and natural systems), seems consistent with Wilson's proposition that urbanity increases desire for zoo visits. Yet, this conclusion may not account for other factors that can impact how social values and cultural norms evolve, including the

---

[16] Manfredo et al. 2003; 2009.   [17] George et al. 2016.   [18] 1993.
[19] Hemsworth and Coleman 2010.

potential role of "educative" nature experiences that Kahn has shown can impact understandings of and relationships with nature in urban contexts.

Sociologist Linda Kalof has demonstrated that social discourses pertaining to animals are nested relationships that often vary by animal type and may not be discrete or fixed. Her findings, which point to both parental transfer of cultural values *and* the role of primarily rural or urban experiences during childhood as potential influences that impact how different cultural groups characterize specific animals, led Kalof to conclude that the dimensions of an individual's value orientations toward wildlife can only be assessed for desired affiliation with consideration of that individual's worldview. Complex and multilayered discourses that reflect concern for animals also highlight an alternative way to consider and explore how biophilia may be at play in the zoo. Rather than assuming affiliation is an evolutionary urge that is narrowly tied to a general desire for proximity to animals, zoogoing could be studied as an opportunity that allows people to reflect on and better articulate or construct how proximity to animals either aligns with or challenges their existing sense of self, their cultural values, and/or their worldview. Such an approach honors the important ways zoos can support zoogoers' explorations of biological relationships and continuity but situates this pursuit as a common meaning-making process consistent with how people of all ages develop a multilayered worldview.[20]

We agree that worldview analysis of experiences in zoos may more clearly illustrate perspectives on the cultural importance of affiliation with living animals, the extent to which and under what conditions zoos appear to increase zoogoers' sense of association with nature and stimulate related meaning-making about the value of animals and their native habitats, and the role of these experiences within the overall social development of zoogoers' commitment to a conservation agenda proposed as zoos' collective mission.

Zoos, in their many historic forms, have always featured narratives and performances of belonging and continuity. The earliest menageries, for example, had the intentional function of reminding those granted the opportunity to explore them that their defined role and place in the hierarchical social structure was grounded in recent and/or ancient history. Trading or gifting animals in the era of menageries expressed colonial power and dominion. Representative of the dominant system of governance, animal collections were used to legitimize and reinforce hierarchy in

---

[20] Kalof 2000.

many diverse contexts. In Oceanic cultures, on the other hand, cats and birds were traditionally situated and presented as companion animals, reinforcing the conceptual interconnectedness of song, sound, affection, and acceptance – rather than expressions of dominion.

The idea that human instinct and environment both influence human attitudes toward animals is, in fact, consistent with Wilson's own updated view on biophilia. A few years after he had initially proposed the biophilia hypothesis, Wilson noted the powerful impacts of learning on our relationships with nature, repositioning humans' instinct to affiliate with nature as a characteristic tempered, enhanced, or neutrally impacted by life learnings.

Social–Emotional Implications of Emotional Connections with Zoo Animals

Past research has suggested that display animals can elicit emotional responses among zoo visitors.[21] Often conflated with feelings, subjective internal states knowable only by oneself and communicated through language, emotions are physical and mental states that drive behavior. Provoked by stimuli, mixed with knowledge, and detectable by involuntary physical responses as well as behavior changes, emotions often provoked in zoos include fear, awe, shame, guilt, and empathy. While complex concerns such as biodiversity loss and environmental degradation seem distant and complicated to many people, encounters with live animals feel personal and may evoke emotional responses that can impact an individual's development of a conservation ethic.

In addition to emotional attachment that connects them to a social network of peers that helps them feel recognized and understood, John Fraser notes that Bronx Zoo volunteers consistently describe being motivated to pursue and continue unpaid work in the zoo by a deep personal desire to be close to and connect with animals. Studies conducted with staff and zoo visitors as part of the WZAM data collection process, furthermore, show that many people who work in and visit the zoo value the site as a natural setting with important restorative qualities, and that compared to members of the general public, zoo visitors have a higher level of emotional connectedness to nature and are more likely to act on their environmental concerns. Part of an unpublished internal report for the Wildlife Conservation Society by the authors of this book, these findings

---

[21] Myers et al. 2004.

formed the basis of the later study of casual zoo visitors' implicit connectedness with nature/natural objects referenced earlier in this chapter.[22]

Attachment to zoo animals may relate to new perceptions or understandings about the biological continuity of human and other animals' sentience or instincts, or to more abstract patterns and conditions of wild nature. Exploring how such attachments are described and realized by zoogoers highlights why living animals in zoos are perceived as an important learning experience for children, and why some adults continue to pursue these social experiences throughout their lives.

Conversations with zoo visitors show that visitors feel connected to the captive animals they encounter,[23] and being near animals in zoos has been shown to provoke perspective taking and a sense of concern for animals, irrespective of the exhibit goals established by zoo designers.[24] Social and behavioral scientists understand the concept of connectedness to nature to refer to people's perceptions about the extent to which they are part of the natural environment. Those who feel strongly connected to nature describe schemas of self and nature with significant overlap, whereas those who do not feel strongly connected to nature describe few points of intersection.

Self-report measures can be used to determine explicit measures of environmental concern and connection, and implicit association tests have been developed to assess unconscious cognitive nature connectedness using stimuli association reaction time to attitudes, self-esteem, and self-concept – with high test-retest reliability and no self-report bias. For the second of the implicit associations test studies highlighted in the biophilia section, for example, which sought to clarify whether zoo experiences have measurable impacts on visitors' explicit and implicit connectedness with nature, and whether such impact is generalizable to a variety of zoo types, data was collected in 2008 from 242 visitors in three types of zoos in New York City. *While implicit connectedness with nature was significantly higher among visitors leaving zoos than among arriving visitors (whether those visitors do or do not arrive with a strong sense of natural connection), self-reported explicit connectedness to nature did not change as a result of spending time in the zoo.* This effect did not appear to be moderated by the specific design strategy or scale of the institutions in the sample, suggesting that non-immersive, naturalistic exhibits can achieve the same degree of implicit association with nature as the more immersive exhibits popular at larger-scale institutions such as the Bronx Zoo and San Diego Wild Animal

[22] Bruni et al. 2008.   [23] Clayton et al. 2009.   [24] Pekarik 2004.

Park.[25] Another important implication? While the desire for explicit connections to nature may motivate some zoogoers, it is deepened implicit connections to nature that appear to be the likely outcome of a zoo visit – though this type of connectedness with nature will not necessarily be apparent to visitors at the time. Interestingly, this valuable, non-recreational, mission-related affordance of zoos may account for (1) some of the enjoyment people associate with zoo visits, and (2) why zoo visitors often have a hard time describing what they find appealing about being near live animals. Future research might explore the extent to which greater implicit connection to nature is sustained after a zoo visit.

Belongings

Museum expectations have been shown to include nostalgic dimensions,[26] and zoo swag and souvenirs frequently become mnemonic devices after a zoo visit, reconnecting past visitors to specific adventures they had in the zoo and to the people and animals that animated those experiences. While critics within and beyond the zoo reasonably point out the hypocrisy of gift shops and on-site vendors selling nonbiodegradable decorative objects that are representative of contemporary human overconsumption of unnecessary, environmentally harmful, rarely used *things*, it is notable that such purchases are not driven by the material or practical value of these items as often as by the desire to acquire a tangible and sentimental demonstration of bonding that can be kept permanently – or at least for a time.

Development of a new tiger exhibition at the Bronx Zoo in the early 2000s involved collaboration with psychologist Valeria Lovelace, an expert in early childhood learning, media, and educational research.[27] Studies with Dr. Lovelace suggested that mothers were particularly interested in tigers and saw these animals as kindred spirits. It seemed that while men in the study sample imagined tigers as primarily male combatants who threatened their masculinity, women felt that tigers were female and most likely quite caring unless they perceived a threat. Women quite literally described having their own inner tiger standing vigil over their children and capable of a deep and unstoppable force if their children seemed

---

[25] Bruni et al. 2008.   [26] Korn 2004.

[27] Dr. Valeria Lovelace is president of Media Transformations, a media research company based in New Jersey. Findings from the studies for the Tiger Mountain exhibit at the Bronx Zoo are detailed in internal reports archived in the Wildlife Conservation Society's library.

threatened. John Fraser notes that while the original intent in the research was to explore whether invoked social stereotypes might support reasoning about animal protection, the retail team became fascinated by the idea of every mom's inner tiger and ultimately produced a variety of products targeted at mothers. By far the most successful item was a simple pair of flip-flops with a tiger pattern; numerous mothers told us this footwear made them feel closer to their own "inner tiger" and that the soft, spongy soles helped them feel they had the strength of a tiger.

When people give or receive a gift item from any distinct setting, they document belonging through the care and consideration of the choice that accompanied the actions of purchase and presentation. Mementos from bonding experiences give us comfort because they reinforce (and sometimes force) memorability, and people often appreciate opportunities to remember shared experiences because such memories can be comforting or evoke a mental warning. In contrast to the zoo-as-chaos metaphor, zoo mementos might reify relational connectedness, wildness, and/or the wider world – and the complexities and continuity of these concepts. And we have established that making abstractions more informed and personal is an outcome of zoo experiences that is valued by many zoogoers, whether that process does or does not fit their personal definitions of learning.

Mihaly Csikszentmihalyi believes all forms of mental flow and most human actions either directly or indirectly depend on memory and memory devices,[28] so we need to be cautious about seeing zoo swag as nothing more than future landfill material. Souvenirs can be simple, silly, and have negative environmental impacts that seem to contradict zoo mission – but they also have symbolic function and often matter in people's lives.

Leveraging Concern as a Point of Connection

Contemporary societies and the communities and individuals that comprise them value and relate to animals and groups of animals in complex and sometimes divergent ways – holding ideas and judgements that can be in direct conflict.[29] Importantly, though, people build on and transform their conceptual knowledge and moral frameworks over time, which is why Peter H. Kahn, Jr. is among the many scholars who note that fostering human relationships with nature requires the embrace of both

---

[28] Csikszentmihalyi 1990.
[29] Herzog and Galvin 1997; Kellert 1997; Kalof 2000; Herzog and Golden 2009.

intellectual inquiry *and* experiential education, "to pay attention not only to nature but to human nature."[30]

Conservation psychologist Gene Myers, for example, has shown that animals impact human understandings of self, human–nonhuman distinctions and dynamics, cultural otherness, and moral functioning.[31] Among the patterns of moral feelings regarding animals that psychologist Gene Myers has found among children is the universal tendency to spontaneously situate animals within their "field of care." Myers notes the importance of awe and wonder as points of entry that foster individuals' emotional bonds with animals and thereby lead to internal tensions because such connectedness inevitably has moral and relational identity implications (such tensions will be revisited in Chapter 9).[32] Kellert has similarly described aesthetic and spiritual appreciation of nature as an openness to "wonder, mystery, and harmony of life in all its vastness and splendor, and the underlying abiotic systems that render this achievement possible."[33]

Awe has come to be understood by psychologists as the emotional response a person can have upon encountering information-rich stimuli that feature physical or implied vastness and overwhelm that person's existing mental frameworks such that existing schemas must be adapted or created anew. Because awe can be elicited by stimuli as wide ranging and subjectively beautiful as art, music, panoramic views, or a natural phenomenon (such as an intricate sea horse), and the emotional function of awe remains unclear, few empirical studies have yet clarified this emotion. Recently, however, psychology researchers have documented a distinct facial expression for awe, determined that awe seems to reinforce social connections, and explored the effects of awe on two social cognitive outcomes: self-awareness and self-concept content. Awe experiences, then, are typically situations that prompt people to disengage from the self and focus their attention on the present conditions making them feel they are in the presence of something greater than themselves and connected to the world around them – outcomes that align well with the process of developing a conservation ethic.[34]

Developmental psychologist Gail Melson describes children as natural ecologists whose engagement with and perceived connections to animals either develop into "respectful kinship, imparting a sense of place in the natural world, a grounded membership in the community of living beings"

---

[30] Kahn 1999, 226.   [31] Myers 2007.   [32] Ibid., 16.   [33] Kellert 2012, 16.
[34] Shiota et al. 2007.

or "weaken and eventually atrophy into indifference."[35] Given that generalization of care-oriented values toward animals and natural environments can grow from children's natural tendencies to care for animals and understandings of animals as subjective others,[36] the zoo can leverage novel experiences that involve awe and direct interactions with multiple species of live animals to help visitors better understand interdependencies and specific factors that can impact animal well-being.

Zoogoers tend to assume that animal behaviors are meaningful and potentially decodable, and naturalistic exhibit design draws attention to the care of animals within the context of their natural habitats and in situations of coexistence with humans. The interactive realms of human–animal relations and the moral complexities of coexistence and care are very much part of the experiential framework of zoo settings. Exposure to and deeper understandings of the fascinating parallels between the minds, relationships, and emotions of animals and humans, for example, can help zoo visitors develop new connections to and understandings about animals. Wildlife Conservation Society research has shown that aquarium visitors are interested in observing animals to understand how they think and are comfortable with a limited amount of challenge to their existing beliefs about human and animal behaviors and needs.[37]

Zoo-based research in recent decades has largely focused on the complex mental states and capacities of a wide assortment of species, with fascinating implications for those tasked with developing educational programming and exhibit design in zoo settings to help visitors feel awe and connection to various animals (and perhaps the biophysical systems that support life).[38] In the mid-2000s, John Fraser was involved in an experimental abstract exhibition without live animals, developed to assess visitor

---

[35] 2001, 189.

[36] Melson's 2001 review of the literature appears to demonstrate that children often perceive themselves to be interdependent with and share a mutual fate with animals, and that when this occurs, children tend to accord animals with duties of care that are very similar to conditions of care and well-being they accord themselves.

[37] Sickler et al. 2006.

[38] For example, animal expressions of elements of culture and language have led some researchers to suspect that comparative research can help researchers clarify humans' unique capacities (see Röska-Hardy and Neumann-Held 2009). In this vein, animal behavior researcher Frans B. M. Waal 2019 has shown that primates pay attention to fairness and experience envy, disrupting the perception that fairness and envy are emotions and behaviors that are uniquely human. Marc Bekoff and Jessica Pierce 2010 have similarly combined behavioral and cognitive research with compelling and emotional anecdotes to demonstrate that animals show moral behaviors such as fairness, empathy, trust, and reciprocity; these researchers go further in their conclusions, though, suggesting that morality is an evolved trait shared by humans and other social mammals.

learning about dolphin cognition in an aquarium. Visitors came away with new awareness of the study of animal cognition and increased understanding about dolphin intelligence. This understanding increased visitors' appreciation for dolphins, as well as their concern for the welfare and conservation of these animals – despite virtually no information in the exhibition about the conservation threats dolphins face. We reported that at least 13 percent of the visitors in our sample were stirred to consider a conservation or concern-related message, even though none were present.[39]

Recent research suggests that the degree of concern people afford to animals appears to be connected to assumptions about and experiences with various animals' consciousness, capacity for emotion and suffering, self-awareness, cognitive capacities, and affection toward humans; as well as the attractiveness of the species.[40] Experiences observing and negotiating such possibilities often involve comparing animals to other animals and to humans and can lead zoo visitors to reassess self-concepts and identity frameworks that sometimes extend to perceptions of kinship. Zoogoers have been shown, for example, to consider themselves in the category of gorillas[41] and affirm kinship with chimpanzees,[42] and while these nonhuman species cannot be accorded full membership in human-designated groupings, verbal negotiation of empathetic concern and social responsibilities to these species has been specifically documented among zoo visitors.[43]

Some highly publicized zoo animals seem to have attained peer status in the United States. Such animals have been memorialized in death, and their personalities and achievements have been described in obituaries run alongside those of [human] community leaders.[44] This honorific treatment shows that the conditions that support such levels of community belonging and emotional attachment are potentially valuable areas of exploration. At the Oregon Zoo, the April 1997 death of a well-known, 55-year-old matriarch elephant, Belle (who had given birth to the first North American captive born elephant calf thirty-five years earlier), resulted in a substantial outpouring of grief from the Portland community. A memorial wake attended by over 1,500 people was staged to aid the community with the public grieving process, and a great degree of debate regarding animal care ensued in the local press and continued to make headline news for years after Belle's death.

---

[39] Sickler et al. 2006.   [40] Herzog and Galvin 1997.   [41] Lukas and Ross 2005.
[42] Bodamer and Sankovic 2000.   [43] Hayward and Rothenberg 2010.   [44] Benbow 2004.

Iconic animals like Belle may be perceived as representatives of community identity and the community's value for nature as represented by animal care practices in its zoos. Collective identification was revealed when attendees joined a ritualized public display of grief and loss, many offering flowers, and cards. This phenomenon suggests that community members who claim an affiliation with their local zoo can develop a common bond tied to the well-being and lives of popular zoo animals that appear to serve as icons for community belonging. Importantly, researchers have shown that when we possess strong emotional attachments to other animals, species, or landscapes, people can become motivated to sustain those entities as expressions of self-interest (rather than altruism).[45] Collective memories of perceived relational value, in these and other examples, ground shared time and place affiliation through emotional connections that emerged through the zoo as a site of collective identity formation with clear environmental ties – and potentially important conservation implications.

Building on previous studies showing that emotional connections with zoo animals can be cultivated among zoogoers during a zoo visit,[46] and can increase visitors' concern for animals' well-being,[47] researchers recently sought to substantiate the belief that emotional connections to a species influence pro-conservation behaviors following a zoo visit. Researchers Jeffrey C. Skibins and Robert B. Powell developed a Conservation Caring scale measuring zoogoers' connection to a species. Determined to be valid and reliable, the Conservation Caring scale is a strong predictor of species-specific post-visit conservation behavior intentions (such as "adopting" an animal) and a weak predictor for biodiversity-oriented post-visit conservation behavior intentions (such as supporting broad-impact sustainability policies).[48] Importantly, zoo visitors in the study sample connected to a wide number of animals, and not all were charismatic megafauna. These findings suggest that while zoo visits are characterized by short-term direct exposure to zoo animals, species-specific on-site emotional connections are a particularly good starting point for the intentional stimulation of pro-conservation behaviors, and numerous species can and should be simultaneously situated as potential flagship species. Campaigns to facilitate conceptual and emotional linkages between flagship species and conservation behaviors that benefit that species, *as well as other species*, seem strategically appropriate, though on-site opportunities

---

[45] See, for example, Kellert 2012.   [46] Myers et al. 2004; Bruni et al. 2008.
[47] Clayton et al. 2009.   [48] Skibins and Powell 2013.

for specific flagship species-focused conservation action taking are likely to achieve the highest levels of visitor engagement.

Human interactions with animals have been shown to support an ethic of care toward specific animals or groupings of animals that can extend to nature more broadly, beyond personal connection or self-interest, which necessarily involves moral consideration of multiple, integrated obligations to animals, humans, and natural systems.[49] In fact, a survey of over 7,000 zoo and aquarium visitors in the United States has recently shown that feeling connected to animals in zoo settings is significantly associated with cognitive and emotional responses to climate change. The research team notes that beyond cultivating zoogoers' sense of place, connection to, and pride in regional ecosystems, zoos may provide opportunities for social support around concern about physically and/or temporally remote problems, even in contexts of perceived political polarization.[50]

On the flip side, education writer David Sobel has shown that many children lose hope and begin to avoid discussing and reflecting and acting on possible solutions when curricula situate environmental problems as unsolvable, widespread, and beyond the geographical and conceptual scope of the learner cohort. Concerned that dissociation and even ecophobia (fear of ecological problems and the natural world) seemed to be unintended consequences of apocalyptic environmental messaging, Sobel became one of the first scholar-practitioners to show that cognitive learning is sparked and the foundation for further empathy development, and potential activism in adolescence can be established when young learners are developmentally supported to identify with and respond empathetically to animals.[51]

The Value of Connections between Zoo Animals and Staff Who Care for Them

As discussed in Chapter 4, a variety of psychological and sociocultural factors lead to the routine devaluation of some species, particularly those treated as pests or food.[52] While livestock transfer venues have been negatively correlated with aversive conditioning (such as pushing or slapping domestic animals) and extremely controversial human behaviors and conditions are common in industrial settings with densely housed livestock, conditions in these facilities have little in common with conditions in accredited zoos and aquariums.[53]

---

[49] Myers and Saunders 2002.   [50] Clayton et al. 2013.   [51] Sobel 1996.
[52] Dhont et al. 2019.   [53] As a starting point, see review in Hemsworth 2003.

National associations such as the AZA, EAZA, and WAZA expect animal care staff to provide continuity and monitor animal behavior. Animal care staff roles, responsibilities, and interactions are inspected as part of the accreditation process. The ratio of keepers to animals is high in accredited institutions, where keepers are specifically tasked with providing enrichment, training, and frequent demonstrations of affection as core aspects of the often long-term relationships they seek to establish with the animals under their care. In accredited zoo and aquarium settings, it thus makes sense to think about animal emotions in the context of intersubjective connections that have reciprocal benefits for the keepers and the animals, and impact what the public can learn about interspecies connections.

Public discourse about zoo animals often includes the question of whether the animals are "happy." Psychologists remind us that "happiness" is only one of many emotional states that animals can experience. As mentioned in Chapter 5, Jaak Panksepp has shown that animals (humans included) experience, express, and learn from a wide range of emotions to develop fitness. Animals in zoos have opportunities to experience the full range of emotional experience – and so do the staff who work with them.

As such, we note that research into animal emotional experience should not be limited to within-species relationships. Unfortunately, though, and for far too long, research on human–animal relationships has overwhelmingly situated animals as "object" and both presupposed and implied a level of generic uniformity in the person-ness of a keeper or stockman's behavior and the response of the studied animal, as if both beings lack individual personalities and operate at the species level in their exchanges. John Fraser has worked with zoo and aquarium staff for many years and describes as abundantly clear the deeply felt and very personal connections between animal care staff and the animals they care for.

The public overwhelmingly bestows upon these professionals the highest level of trust, in no small part because while visiting the zoo they often have opportunities to witness authentic connections firsthand.[54]

Among the 130 zoo professionals who participated in a study conducted in the United Kingdom by Geoff Hosey and Vicki Melfi in 2012, 103 claimed to have established bonds with animals, approximately three-quarters with zoo animals in their care. These results were most common among staff working with primates and carnivores. Reflecting on why such connections are beneficial, study participants shared that these

---

[54] Fraser and Sickler 2008b.

bonds make their work easier and produce affective benefits in terms of their sense of well-being and job satisfaction.[55] Hosey and Melfi have continued this line of inquiry, suggesting that the value of these emotional connections are the foundation for advancing understandings about how the psychological experience of connection might flow beyond direct interaction moments.[56] At least one zoo-based study of chimpanzees and visitors has identified visitor–chimpanzee connections as intersubjective exchanges that are acted upon in ways that are satisfying for both participants.[57]

In zoos and aquariums, the potential benefits of these relationship investments suggest that we should assess the temporal nature of keeper schedules, looking at, among other foci, the impacts of predictability and change. Whether regular schedules become disrupted when a favored keeper takes a two-week vacation or is transferred to another area for some duration, for example, remains understudied. It may be that these types of temporal changes in relationship predictability contribute positively to the development of healthy ways of coping with loss and change, for some species or individual animals, but have negative impacts on others.

Researchers exploring the unknowns of emotional enrichment should think strategically about how to assess the emotional relationship embedded in being a social human working closely with nonhuman animals as part of the daily work of animal keeping. Are there tools and techniques for helping animal care staff think about emotional or potential emotional transfers (to fellow keepers or to animals), in order to leverage those experiences as enrichment opportunities; document how that experience is part of the zoo experience; and develop protocols that can ensure these practices are in the best interest of the animals in our care?

Any study that attempts to describe what contributes to the full range of emotional experience and learning should likely include some way of characterizing how management decisions, staffing strategies, and temporal change are variables in the study. A day that a keeper might describe as emotionally challenging may not necessarily have long-term negative consequences on an animal – negative interactions are part of living a full emotional life. Unless we track and study variation over the long term, though, we cannot really account for this issue and its impact. While this may not be a big issue for animals whose natural lives are brief, many zoo species have lives in one institution that span the careers of multiple "long-term" care staff.

[55] Hosey and Melfi 2012.   [56] Ibid.   [57] Persson et al. 2018.

Considering these variables leads us to wonder whether animal care professionals might be more self-aware supporters of improved animal welfare if they attend to their own emotions as a factor in the emotional life of an animal with whom they interact. Studies have demonstrated that human well-being is improved through the presence of pet animals as part of human domestic life and the exercise of caregiving to these animals. Anecdotally, it is community knowledge in the zoo community that some devoted keepers feel their job satisfaction is compromised when they are required to miss a day with the animals for work-related reasons, and zoo veterinarians rely on keeper reports of an animal having an "off-day" as an indicator that the animal might need attention. However, we have yet to operationalize what this means as part of a much larger body of literature on interspecies emotional experience, regulation, and management.

Visitor studies have demonstrated that keepers are accorded special status based on a perceived positive emotional bond with the animals in their care, but researchers have not really explored this construct or what supporting a full range of emotional states in zoo animals might mean to the advancement of conservation learning at zoos. As has been the case with recent research that points to a complex array of human thoughts about animal thinking, commensurate research on the complex emotional relationships between those who work in zoos and aquariums and the animals in their care might contribute new dimensions to zoo interpretive strategies to advance public understandings of the work of zoos and the implications of environmental change on the ability of animals to live full, emotionally rich lives in the wild.

From Connection and Concern to a Shifting Moral Compass

Instigated by the federal administration at the turn of the decade in 2010, the National Science Foundation undertook an ambitious grant program to fund large national experiments that sought to help increase public understanding of the science behind climate change. One of those grants was awarded to the Chicago Zoological Society and another to the New England Aquarium. Both projects involved multiple partner institutions collaborating to understand the affordances of zoos and aquariums to increase public engagement with climate concepts and issues. Led by Alejandro Grajal, the Chicago experiment involved a survey of zoo and aquarium visitors in numerous regions of the United States. The research team found that the sense of being connected to animals was predictive of pro-environmental behaviors in general, but also specifically predictive of self-efficacy in tackling the issue of climate change. Importantly, and to the surprise of many, they found that *political inclination on the conservative to liberal spectrum had no impact on visitors' sense of connection to animals and the resulting pro-environmental stance.* These findings demonstrated that, for zoogoers, valuing human–animal connection is a nonpartisan issue and is linked to pro-environmental action, indicating a psychographic variation between zoogoers and the general voting public overall and pointing to the social value zoos offer as experiential pathways to conservation-focused behavior change.[58]

In this chapter we have shown that live animal observation opportunities often challenge people's established frames of reference and provoke positive emotional reactions such as awe, respect, and appreciation of natural beauty. Moreover, when zoo animals appear to pay attention to them or they observe zoo animals exploring what they perceived to be a shared world, visitors tend to describe an emotional connection that increases their desire to help those animal types.[59] John Fraser notes that children, for example, typically focus significant attention on extending justice and care to zoo animals by learning their names, imagining ideal conditions for those animals, and understanding how to care for them. Many are also very eager to enact, in the months to come, new ideas about quality caregiving with a new animal toy purchased at the zoo gift shop.

Since the mid-1990s, zoo professionals and analysts from a variety of fields have been making the claim that exposure to living animals "develops empathy, which translates into conservation action."[60] Data continues to accumulate that the empathy development link is solid. Quantitative and qualitative data show that elementary classroom teachers who use live animals as classroom pet teaching tools believe this opportunity increases students' empathy and social–emotional development.[61] Conservation psychologists have established that perceived similarity or phylogenetic relatedness to animals is associated with higher levels of human empathy and significantly correlated with the desire to help care for those animals,[62] and that empathy toward animals can impact individuals' capacity to extend their scope of care to nonhuman entities in important ways that can sometimes override cultural norms.[63]

---

[58] Grajal et al. 2017. We describe the New England Aquarium's collaborative project, led by William Spitzer, in Chapter 11.

[59] Myers et al. 2010.   [60] Norton et al. 1995, 327.   [61] Daly and Suggs 2010.

[62] Sickler et al. 2006; Clayton et al. 2009.   [63] Clayton and Myers 2009.

A group of zoos and aquariums in the Pacific Northwest has been working since 2015 on a long-term collaborative study to understand the value of empathy in promoting conservation. The researchers are pursuing this study with conscious awareness of critiques that zoos may actually be (structurally) predisposed to the diminishment of empathy,[64] though their evidence to date suggests that interactions at the zoo do sometimes enhance zoo users' perspective taking and understandings of an animal's needs and life patterns. The research team remains dedicated to this ongoing research based on the shared belief that empathy processes can help zoos improve their efficacy for promoting behaviors that lead to the protection of biodiversity. This research promises to shed new light, in the coming years, on empathy as a key aspect of the functional role of zoos in human development.[65]

Given the data pointing to the wide range of positive outcomes that tend to accrue as a result of human–animal/nature relationships, some scholars suggest that zoos and zoo programming should focus much more heavily on choreographed experiences of care and concern in order to facilitate emotional connections to animals and caregiving that build on and inform zoogoers' existing ideas, concerns, and motivations.[66] Increasingly aware that they have opportunities to foster among visitors a sense of awe about or shared identity with various animals in their collections, many zoos now offer adopt-an-animal programs that situate specific animals as part of visitor families (notwithstanding the fact that the adopted animal will never leave the zoo setting), as a pathway toward deeper perceived connections. In a similar vein, exhibit designs and signage in many zoos has been recently reimagined and redesigned to encourage zoogoers to identify similarities between themselves and zoo animals through explicit attention to shared (and potentially observable) physical, motivational, or behavioral patterns.

A range of tangible benefits appear to accrue to zoos and zoogoers alike when exhibits combine animal training with interpretation to highlight animals' needs and special abilities and show how people and animals can form bonds and work together. Comparisons of zoo visitors who experience zoo staff offering a live animal demonstration and zoo visitors who passively view those same animals and read signage in the same exhibit setting show that the former cohort have extremely high content and

---

[64] See Chapter 4 regarding the arguments of Beardsworth and Bryman 2001, Acampora 2005, and Kahn et al. 2008.
[65] To access a description of the project, see www.zoo.org/empathy.   [66] Rabb and Saunders 2006.

concept retention for weeks after their zoo visit (education benefit).[67] Visitors who view a live animal demonstration with staff interpretation also express greater support for conservation efforts than visitors who view the same animals without staff interpretation (mission-supporting benefit).[68] In addition, live animal demonstrations complemented by oral interpretation have been found to be entertaining (recreational benefit),[69] increase visitor perceptions of the zoo, and significantly increase mean exhibit stay time.[70]

These patterns have clear alignment with the goals of the modern zoo – research, recreation, education, and conservation – yet information alone is rarely enough to promote pro-environmental behavior.

Even empathy for animals is unlikely to lead people to develop and act on a conservation ethic unless this emotional response pattern is linked to a moral problem that feels important and potentially solvable through new attitudes and actions. We will continue to explore various intersections of empathy, care, moral development, identity, and action in Chapters 9–11, but note that knowledge and empathy will need to be complemented by motivation, for conservation goals to be realized. It is thus encouraging that there is reason to believe that perceived connections with one or more animal in the zoo can be an emotional factor that provides or supports such motivation for some zoo users.

---

[67] Heinrich and Birney 1992.   [68] Swanagan 2000; Anderson et al. 2003.
[69] Yerke and Burns 1991.   [70] Anderson et al. 2003.

CHAPTER 9

Identity
Discovering Self

"Please. *Don't* tap on the glass." A perpetual problem at zoos and aquariums, this behavior is so widespread that it must have a psychological root.

John Fraser once did a study formally observing people at a rhinoceros exhibit at the Chicago Zoological Society's Brookfield Zoo. A father and son of about seven years old walked up to a viewing area, but the rhino did not look up from the forage she was messing around with in the middle of her yard. She was alone but seemed quite content to meander around to smell and discover whatever had happened there a short time ago. The dad called out to the rhino using phrasing we now know is important. With his left arm on his son's shoulder, he called out "Hey rhino, hey there! You. Hey ... hello ... hey." He then reached forward and pulled a small plant out of the earth in front of the exhibit and held it out to the rhino, yelling more loudly, "Hey rhino, we're OVER HERE," while shaking now-loose branches. Alas, the rhino would have none of it.

Clearly, zoos would be devoid of plantings if every visitor ripped them from the ground, so this aspect of the situation was rare behavior, but the situation overall was not as uncommon as we might like to imagine. It's not that people want to harm animals or make their lives unpleasant, but those tasked with the care of these animals know this behavior is bad for the animals – and deterrent signage is typically abundant. Actions to try to coerce desired responses from zoo animals are so predictable that the pattern frequently finds a way into comedy program content. Windows are particularly problematic because zoogoers know they are two-way devices, not TV screens where the focal entities in the exhibit can't see the actions of the audience.

Conservation psychology pioneer Carol Saunders found this need for personalized interaction with animals at the zoo a fascinating phenomenon. In addition to exploring the range of emotional experiences that unfold when people spend time in accredited zoos, Saunders was interested in the unique value of experiences where an animal is perceived to be "attending" to a visitor by looking at or otherwise interacting with the individual in some way that suggests this animal has validated that person's presence.

Even those who work in zoo settings find these types of experiences meaningful and memorable. Alone before opening hours, John remembers inspecting a recent sign upgrade and noticing a grouper that had nibbled a bit of food and was goofing around in the tank. He also immediately noticed that the fish was gazing directly at him, though nothing about what John was wearing or carrying indicated he might be a source of food. When their eyes met, John had immediate wonder about the lifeworld in the tank. This fish undoubtedly knew there was a barrier and that stuff happened in the terrestrial world that impacted conditions within the tank. How much of that knowledge was based on vigilance and observation of change? What seemed certain was that the grouper had its eyes on John, a being, not an object but a subject doing things that might require a response or impact options, a subject to monitor. These moments of intersubjective exchange left John feeling he knew that grouper as a thinking and self-aware individual, and that grouper had its eyes on John, too. In fact, its eyes, on John's eyes, looked away and then recon-nected. They saw one another.

In 2004, Saunders and colleagues found that such moments of being seen and acknowledged by zoo animals were occurrences zoogoers find highly satisfying, thrilling, and fulfilling.[1] From a psychological perspec-tive, such an occurrence is understood to be an intersubjective exchange, two or more beings (e.g. human and grouper) simultaneously recognizing each other's distinct presence at a specific moment of time. Applying this concept to the situation of the father and son observed at the rhino exhibit, four subjects were involved in the exchange described. All saw a diffident rhino ignoring John and the pair standing on the other side of a protective moat. John and the son were both very aware that a father on the scene was aggressively soliciting an animal interaction for his son but not noticing a lone, middle-aged man with a camera and clipboard taking this all in. John and the father were both aware of a son being held affectionately by his father, but only the son was observing John observing their behaviors. Based on that boy's rather sheepish downcast eyes, the son also likely recognized that uprooting plants in the zoo is transgressive behavior that could have negative repercussions, though such fallout was unlikely to be anything he had the power to alter. Though we cannot know what the

[1] Myers et al. 2004.

rhino was thinking and perceiving, all four subjects were part of a social exchange at that moment. The rhino sensed our presence, she just found us irrelevant and was thus inclined to ignore us.

Previous chapters have shown that human social responsiveness and responses to animals are deeply connected to cognitive, emotional, and moral development. Human expressions of care toward animals and nature are thus inevitably rooted in social development and sense of self in relation to other beings.[2] This chapter explores the psychological under-pinnings and importance of identity and identity work to zoo experiences and zoo mission, situating identity as the fluid understanding of who we are and how we choose to operate in society, and identity work, which occurs through many mechanisms, including play, as effortful and con-scious experimentation with self-presentation and the solicitation of feed-back from others.

Identity Work

American poet Gertrude Stein published *Identity a Poem* in 1935, a few short years before the eve of America's involvement in the Second World War and the German invasion of the lively Parisian cafe life she had embraced as her second home. As the world as she knew it began to be overcome with explorations of national identity, xenophobia, and hate speech that threated both her Jewish heritage and her queer self, Stein paused to look introspectively at how a person comes to know the limits that encompass one of the central psychological questions of her century: the concept of knowing one's self. Ultimately developing a genre- and gender-defying book-length poem, play, essay, she iterated between how we name, perceive, and define this type of knowledge. The opening line of Play 1 reads: "I am I because my little dog knows me. The figure wanders on alone."[3]

How concisely and provocatively Stein introduces human internal strivings to understand and act out who we are! The process is inevitably complex. An individual may pridefully announce her cultural heritage as a source of behaviors and beliefs, but cringe and feel despair when subjected to biased attacks. All such expressions involve identity work. Stein wrestles

---

[2] We note that this same reasoning underpins Myers and Saunders' 2002 view of biophilia as a byproduct of human *social* evolution, rather than an adaptive preference for biotic environmental features.

[3] Stein 1993.

in her play with language about self-understanding regarding her full being, suggesting, to initiate that process, that a little dog, her pet, might know more about her than she or other people do. As Stein's poem iterates the many ways people play with their position in society and the feedback they receive, she describes the ever-evolving work of self-understanding.

Deepening understandings of the self and identity-related processes are pursuits foundational to the discipline of psychology, which was initially conceived as the study of how mental processes work.

The self is our internal understanding of who we are. People happily create labels for themselves as animal people, science people, and environmentalists, but they are also parents and children, leaders or team players. They come to wish or know these fluid features of themselves over their lifetimes.

Roy Baumeister's 1999 book *The Self in Social Psychology* is a collection of articles outlining several decades of the important studies initiated to understand how people come to understand who they are. Recognizing that intersubjective experiences represent feedback from the world around us that provide crucial data points for "self"-study, Baumeister notes that three feedback pathways dominate the process of self-understanding.[4] People understand the physical self through what they can touch or feel as a sensation (proprioceptive or somatic feedback). Most psychology studies of the intersubjective have explored what others say to us or what we hear, which is the second feedback pathway. Yet, those who study animal interaction and computer interfaces readily acknowledge that feedback in these domains can also drive beliefs about who people believes themselves to be. The third pathway Baumeister focuses on relates to executive function, the mental processes people use to engage in a behavior to produce a predicted reaction. It is, he argues, the ever-evolving interplay between these domains that helps a person come to understand their self or sense of self.

Researchers have long sought to understand how contingencies and settings influence our identity work across the life course. A child at the zoo might be involved in identity work while imagining herself as a superhero who will save all the lorikeets in the world, while her grandfather is reflecting on and enacting who he wants to be in that child's life and how to position *as a grandparent* in relation to environmental topics on every sign that surrounds them. Such self-work likely involves that

---

[4] Focal topics include self-knowledge, self-esteem, self-regulation, self-presentations, and the self and culture. Baumeister 1999.

grandparent negotiating a wide range of identities, including, perhaps, passionate hobby pursuits and who he was at work before retirement, and comparative memories of himself at the current age of this grandchild.

Falk and colleagues became quite intrigued by the idea of identity work in their studies of learning at zoos and aquariums.[5] As described in Chapter 3, their findings revealed five dominant motivations for choosing to visit a zoo or aquarium, with novelty, hobbies, and interest in facilitating experiences for others as three of the five dominant modes of engagement. While "Facilitator," "Experience Seeker," and "Professional/ Hobbyist" motivations are not identities, each of these behaviors is implicated in identity work. This is inevitable because various behaviors that reflect the "selves" that individuals with these motivations present in the context of zoogoing become part of the social experience of the visit and have implications for the types of engagement and feedback they seek on site.

With identity work, a person engages in acceptance-seeking behaviors from those whose opinions they value. When we introduce this topic, people typically think about conditions like narcissistic personality disorder, which is an extreme form of self-focused attention seeking. While the causes of narcissism remain unknown, current consensus suggests that this condition likely involves a combination of genetic and environmental factors that result in an excessive need for admiration, disregard for others' feelings, and a sense of entitlement, among other symptoms. Most people, in contrast, hold some degree of altruistic values, and their identity work is a constant exercise in contingency testing to understand the degree to which they can and should extend their scope of justice.

To bring this back to the zoo, when we hear that people "care" about animals or nature, we understand that personal experiences at a zoo are extensions of these individuals' seeking behaviors. The identity work they do at the zoo is personal social exchange that helps them understand the degree to which their sense of altruism extends. Identity work is quite distinct from an egoistic focus. Identity work is a social process, and zoos serve a vital function in that process. The zoo facility is a grand presentation of novel conditions that allow the people therein to play with who

---

[5] In 2008, Falk et al. proposed that that the temporal strivings of visitors at the zoo could be grouped into five dominant motivations. That work was somewhat contentious, and in later studies outside zoo and aquarium settings these authors softened their initial terminology, describing the patterns they sought to describe as "identity-related motivations."

they are and what they're about. Zoogoers can elicit feedback from their loved ones and come to a deeper awareness of who they seek to be.

People engage in identity work when they choose a behavior and pay attention to the feedback they receive. When parents situate and present themselves as Facilitators using the zoo as a place to instill moral values in the children they bring with them, they use on-site experiences to "do" identity work by moving, talking, and learning together with their children to develop shared understandings of a novel setting. In doing this work, they also set up conditions that help them feel accepted within their group.

Psychologist and professor of human development and social policy Dan McAdams has focused on how we reconcile the many identities we negotiate in our lives, particularly in different social settings. Effectively, his research demonstrates that a person will draw on their multiple identities in different social settings, electing to prioritize as needed. This is not to suggest that these identities are in conflict, but they are tested in a zoo setting, where a person will iterate between their identity as a parent, a nature lover, a child, and someone who is afraid of snakes, but has fond memories of the feeling of companionship they received from their little dog in childhood. The study of relationships between identity and zoo experiences thus encompasses many angles and perspectives, and visitors and communities define identity in relation to the zoo and its mission in a great many ways.

Zoos: Unique Sites for "Self"-Learning

John's brother-in-law visited the zoo when he and his wife went on their first "serious" date. Actually, many people bring with them to the zoo a person or group of people they are trying to know better, which makes sense because the stimulus of seeing living animals in idealized settings and idealized groupings elicits behaviors and discussions that help people playfully sort out, articulate, and present who they are.

Zoos, and museums more broadly, allow for free choice learning. From the perspective of identity work, learning choices are not related to the institutional agenda for teaching, whether that's science or conservation, or simple animal facts. Engaging in identity work is the effort to question WHO I AM in relation to the stimuli surrounding me. That all-encompassing surround includes learning who I am in relation to others in my social group. If one person is afraid of a snake, I will need to draw from my multiple identities and experiment with how to respond to their fear. I may focus on my understanding or knowledge to share information as a way to dispel their fear, laugh, and giggle because the zoo creates a safe place for us to negotiate the meaning and history of that fear to build a stronger bond of us as a pair. Essentially, the dialogue and our behaviors become identity work that has not been tested until that moment.

The social exchanges of the visiting group can thus be understood as identity work for the members of the group. The novelty and need to understand themselves become the primary points of reference for accepting or rejecting any learning pedagogies imposed or introduced by the institution. Indeed, there is reason to believe much of the appeal of museums and museum-like experiences rests in the ability of such venues to offer visitors opportunities to experiment with alternative identities, given that museum visitors have been shown to explore contingent relationships and new models for self-expansion that can be stored in the memory until they are needed.[6]

Deeper understandings of zoos as social and physical environments that impact identity and inform how various groups negotiate their concern for captive wild animals affords scholars, conservation psychologists, and zoo staff alike the opportunity to better understand the role animals may play in the psychological development of social concern for animals and the environment more broadly. Indeed, connections between concern for animals, self-concept, and social norms regarding wildlife allow us to consider how animal relationships inform personal and collective consciousness and decision-making. Three aspects of identity work that distinguish zoos from other cultural venues seem appropriate to highlight here.

The zoo's presentation of numerous undomesticated (and some domesticated) live animals create a concentrated range of stimuli for exploration. These presentations introduce the concept of natural living systems of which one is a member, but at a scale and complexity that can be interpreted as chaos (see Chapters 4 and 6 regarding zoo–metaphor dynamics). Various metaphoric constructs thus become available for contingency testing in novel on-site scenarios, often pointing to how one's identity is distinct or maps to identities observed in other living beings. When we overhear a mother explaining to her children that "that Mommy gorilla is protecting her children from the bully," we recognize that the gorilla may have no such motivation for intervening in this emergent activity in this zoo habitat, but can be sure that the mother uttering these words is actively engaging in identity work. She is situating and projecting

---

[6] Carbonell 2004; Rounds 2006.

her own identity into an immediate situation and seeking affirmation from her children regarding their relationship. Such an effort is enjoyable because it is likely to affirm the respective social identities of her visiting group and align with her motivation to facilitate recognition of relationships and risk.

The zoo-simulated "natural" environment, a microcosm of some aspect of the planet, also offers a conceptual frame that helps focus zoogoers' identity work on scale, responsibility, and victimhood. It is the zoo's intention that zoogoers understand themselves to be part of a much larger human enterprise that interacts with nature and need to be challenged to define our understandings of aesthetics, science, systems, and relationships. As described in Chapter 8, the animals and plants are part of zoogoers' desire for and sense of continuity. Animals serve as one bridge to caring about the natural world, according to many theorists, not only because nature (and animals as part of nature) is a source of socially constructed symbolism from which people derive meaning, but because nature and animals are highly and enduringly compelling to humans of all ages, eras, and cultures. Dynamic and highly responsive, animals have particular appeal to humans due to human propensities to engage socially and are the only nonhuman entities that allow for direct experiences with living, feeling "others" – conditions that have been shown to be integral to sense of self during childhood and often lead to care-focused outward actions.[7]

Zoos are also crowded places where large numbers of people are simultaneously pursuing their own identity work in a close-proximity social context. The Jerusalem Zoo, for example, is a particularly interesting venue because historically segregated cultural groups find themselves standing side by side observing animals. Individuals visiting with their Muslim, Jewish, or Christian family explore their personal identities in relation to the animals and natural world while also navigating and negotiating, among other layers of identity, their faith, reflected in sometimes divergent norms and expectations around social propriety.

Place Identity

Data related to the knowledge of how human social abilities develop and are connected to both physical environment and action underpin the social interactional perspective of many conservation psychologists. Human interactions with the physical environment may impact self-concept.

---

[7] Stern 1985; Melson 2001.

While *sense of place* speaks to unique, external attributes of a location that help individuals situate themselves physically and socially within that specific context, *place attachment* is a multidimensional concept related to "embeddedness," an emotional bond with a known physical setting that can arise when an individual's place-based experiences intersect with that individual's feelings, memories, and perceptions.

Taking psychological studies of place in a different direction, in 1978 environmental psychologist Harold Proshansky proposed that *place identity* can be a key aspect of self-concept. Attending to the located nature of subjectivity, Proshansky sought to identify whether and how positive experiences with place-based cultural institutions might contribute to urban children's positive self-concept, challenging, as he did so, conventional assumptions that cities were socially detrimental and combative settings, and highlighting the important socialization role of cultural institutions.[8]

While "nature" has remained a somewhat loosely defined construct in the four decades since, conservation psychologists, researchers, and theorists from multiple disciplines have continued exploring how the presence of natural living elements, and changes to those living elements over time, might be connected to a concept of self that reflects integral connection with nature.[9] Place identity has been shown to feature an emotional and symbolic affective dimension and psychometric properties of place attachment that inform the meaning an individual ascribes to a specific, known place.[10] Showing that place-based experiences, over time, can impact how individuals construct an environmental identity, for example, sustainability scholar Mitchell Thomashow has demonstrated that more robust relationships with nature can emerge when individuals combine scientific learning and perceptual explorations of nature.[11]

Proshansky's original theory recognized that constructed experiences at cultural institutions socialize children into the dominant local culture and play a significant role in the development of place identity components of the self. As described in the opening chapters of this book, zoos are enduringly popular cultural institutions recognized as a unique museum type that allows communities to interact and connect with dominant cultural views of wild nature and the living world. There is evidence to suggest that in the late twentieth century Western museums evolved from

---

[8] Proshansky 1978; Proshansky et al. 1983.
[9] Bott et al. 2003; Clayton and Opotow 2003; Kahn and Kellert 2003.
[10] Manzo 2003; Kyle et al. 2005.   [11] Thomashow 1995, 2002.

their original, material-culture-collecting function, transitioning into a form of mediating environment wherein community members can physically explore shared stories; openly negotiate, together, their understandings of and relationships to common ideas; and make meaning around common issues.[12] Can zoos, then, be conceived as monuments to the dominant culture's views on how animals are valued and how such values are central to community-level place identity?

Well, news media often report zoo conditions and operating norms when they seek to describe or allude to community well-being, particularly in extreme natural or human disasters situations. The reopening of the Audubon Zoo in New Orleans, for example, was explicitly framed as an implicit indicator of community-level recovery following the extreme devastation wrought by hurricanes Katrina and Rita in 2005.[13] In a similar vein, journalists and media outlets highlighted local perceptions of the depths of human suffering during wartime by focusing on the sacrifices of those caring for various animals that had been abandoned at the Baghdad Zoo during the 2003 invasion of Iraq by the United States.[14] Indeed, consistent graphic descriptions of citizens' emotional relationships to their local zoo and local zoo animals do situate zoos as local institutions central to individual- and community-level sense of place, place attachment, and place identity.

As John Berger points out in his critique of zoos, the place identity contribution of zoos may be more a mnemonic function connected to humans' desire to organize nature according to community priorities than any actual representation of nature.[15] While the Bronx Zoo, Singapore Zoo, San Diego Zoo, and Smithsonian's National Zoo in Washington, DC are world-renowned institutions that receive national funding, and may represent dominant national views, most zoos receive local-level government funding, manage their own advertising, and are marketed as tourist attractions unique to the locality – though zoos can feel (and often are) remarkably similar in design and scope. In practice, then, zoos may be more bound by their civic role rather than the supra-national conservation agenda mission focus of their associations, which is likely why zoo experiences overall have been shown to reflect localized relationships to the culture of nature.[16]

Research suggests that while place identity can reinforce existing perspectives, moral values connected to place identity can also be challenged

---

[12] Archibald 2004.   [13] Associated Press 2005.   [14] Dahlburg 1996; Bishop 2004.
[15] Berger 1980.   [16] Swanagan 2000.

and contradicted.[17] John Fraser has demonstrated that groups protesting evolution exhibits in the United States not only had views that diverged from those of the zoo and science museums curation and management teams that introduced the exhibits, but also perceived the exhibit content as a significant violation of many visitors' place attachment and sense of belonging to the host institution.[18] Animal rights protests and sociocultural narratives about zoo outcomes connected to war and natural disasters similarly point to zoos as high-profile situated monuments at controversial intersections of culture and nature.

These patterns highlight the status of zoos as active civic spaces wherein sociocultural actors accord concern, confer meaning, and assign authority for representing how to appropriately view and care for animals and nature. Beyond supporting personal experiences of place attachment and explorations connected to self-concept precursors that appear to ground environmental identity development characterized by concern for animals and natural systems, zoos are sites and sources of public debate regarding the social and moral values and roles of various groups within and beyond the local community through lenses of place identity.

Human Responsibility: Environmental Identity Development

John Fraser focuses on how to promote conservation values through engagement with place and emotion.[19] Interviewing families at the Philadelphia Zoo to learn about why they were visiting and determine what and how they learned from a new exhibition, John and his colleagues were also interested in gathering data on these visitors' environmental values. One family was particularly memorable. Not only had they renovated their house to live off the grid, but their diet, purchases, and occupations were viewed as opportunities to be intentional about reducing their impact and repairing the planet. Even so, the mother explained that she brought her children to the zoo so they could learn as a family what more they could be doing to live sustainably. She saw the zoo as a critical ally in her effort to live and model a sustainability ethic.

Environmental identity, which this family so clearly epitomized, has been described as situating the self within the natural world, evident through the degree to which care and concern normally accorded to other people is extended to nonhuman life.[20] Systemically acting as if elements

[17] Dixon and Durrheim 2000.   [18] Fraser 2006.   [19] Fraser and Brandt 2013.
[20] Thomashow 1995.

of nature are part of the self has been defined as the exemplification of personal ecological identity,[21] and has been shown to be a measurable element of self-concept.[22] Environmental identity, thus conceived, is a bond to nature that often emerges through contextual experiences related to the social process of building one's sense of belonging with others, including other animals.

Susan Clayton has developed an environmental identity scale to assess personal identification with nature based on scales and constructs that have been developed to study collective identification with other people. Clayton's instrument specifically focused on salience of the identification with nature, inclusion of self as part of nature, agreement with or support for an ideology that associates the self with nature (such as environmental education programming), and positive emotional feedback associated with being in nature. This environmental identity scale considers nature as a functional collective identity reflected by behaviors related to shared concern that are attached to the individual's relationship with one or more environmental entity, as opposed to a group or groups of people.[23]

Zoos As Sites Where Religious and Environmental Identities Overlap

All world religions share the sense that humanity is connected to a universe of underlying order, with ethical implications for our connectedness to the other creatures that inhabit our world.

In contrast to the social motivations that characterize most zoogoers, WZAM researchers found that Spiritual Pilgrims are a visitor grouping that selects zoo experiences for more introspective purposes.

Fairfield Osborn described "spiritual concepts" intertwined with the relationships between people and natural resources.[24] St. Francis argued that humans and animals are co-worshipers. Aldo Leopold emphasized the connections between humans' spiritual health and wilderness preservation.[25] With often convergent perspectives regarding responsibilities to future generations, Leopold, John Muir, and Henry Thoreau, challenged dominant perspectives regarding humans' relational and spiritual kinship with – and moral obligations to – the natural world.[26]

---

[21] Zavetoski 2003.   [22] Clayton 2003.   [23] Ibid.   [24] Osborn 1953, 4.   [25] Nash 1982.
[26] David Kinsley 1995 identifies Muir, Thoreau, and Leopold as some of history's most persuasive and influential voices for ecological spirituality in his cross-cultural, historical exploration of humans' consistent, intuitive sense of being part of a moral order that extends beyond other humans. Kinsley presents numerous philosophical, theoretical, moral, ethical, and spiritual constructs that situate

In 1999, a national market survey of 1,000 adults in the United States revealed that 75 percent of respondents agreed that visiting a zoo brings them closer to nature – and 75 percent agreed that zoo visiting has brought them closer to God.[27] These results point to very high, and often simultaneous, levels of environmental and spiritual value accorded to zoo visits, and we note that, like the proposed phenomenon of biophilia, religious doctrine situates individuals within the context of the natural world and continuities therein.[28] Particularly in times of crisis, the conviction that life is purposeful, connected, and coherent gives people comfort. Perhaps this explains why colleagues at zoos and aquariums (and botanical gardens and nature centers) across the United States describe similar stories about attendance spiking after 9/11, with visitors coming not for programming, per se, but to contemplate.

Many rabbis teach that appreciating the world that God created for man's benefit necessarily requires taking care of that world. In addition to asserting that all of nature represents the physical manifestation of the law of God in chapter 2 of the *Mishneh Torah*, Maimonides suggests that reflecting on the wonders of creation leads to the spontaneous love of – and desire to try to comprehend – God. Aligned with this teaching is the Jewish community practice of zoogoing, which has long been presented as an ideal family experience during the holidays. Members of Hasidic communities visit the New York Zoos and Aquarium in large numbers during Chol HaMoed, the intermediate days of Passover and Sukkot, seeking shared experiences of wonder and intergenerational identity formation in the context of the stimulus of living animals and in the presence of others who are visibly recognizable as being part of their faith-based community encountered during the visit, performing parallel ritualized family activities.

WZAM researchers interviewed twelve leaders from a variety of faith traditions to explore the perceived connections, conflicts, value, and

---

ecology and religion as deeply connected, noting that practices of cultivating rapport with animals often emerge at these points of convergence.

[27] This was a proprietary market survey; John Fraser was personally involved in design, implementation, and analysis.

[28] Christian, Buddhist, and Taoist traditions, for example, all recognize humility as an esteemed virtue appropriate because of human smallness within a larger system, and followers of those faiths then try to make sense of and put into action what form that knowledge might take as a conscious practice. Judeo-Christian sacred texts are replete with nature metaphors that speak to affinity for God's work. The dove is understood to represent peace, new beginnings, and, for Christians, the Holy Spirit. The rainbow and the olive tree are understood to represent covenant promises. Islam teaches adherents that cats are creatures that should and must be cherished and loved, and Eastern religions situate animals as givers of life.

relevance of zoos and aquariums to spiritual and faith-based communities. Respondents consistently described spiritual and religious narratives as sociocultural tools that help people make sense of the diversity and complexity of natural and human systems. Overall, these spiritual leaders described valuing zoos for reasons connected to accessing nature and animals, and the ways they spoke about the importance of humans taking responsibility for ecological well-being sounded very much like the narratives of zoo staff interviewed in the same study series. Both groups tended to focus on how to be optimally effective about using zoo animals for illustration, education, and, in some cases, modeling principles of care. Evolution, which the research team expected to be a contentious issue, was never raised by any of the leaders in the sample as a zoo-related concern; the personal interpretations they shared seemed to allow for the compatibility of intelligent design and evolution concepts.[29]

While underlying theological goals that affirm caring for animals and nature as part of religious practice may be more doctrinaire than the shared goals or affinities of other groups, these examples make clear that zoo animals serve as resources for rich discussion and morally grounded identity development, with potential conservation implications, rather than simple entertainment. These findings point to opportunities for programming that explores and tests the potential power of spiritual pursuits as a motivation that has individual and group-level moral implications that might overlap with, and thus spark or reinforce, the environmental identity of some zoogoers. An example of the intentional use of ritualized zoo visits to establish parameters and limits around in-group religious beliefs and identity formation, privately administered "Biblically correct" tours of the Denver and Colorado Springs zoos have offered fundamentalist Christians and homeschoolers creationist, young-earth exhibit interpretations that frame speciation, complexity, and morphology in completely nonmainstream ways. Though extremely controversial, this type of unscientific alternative messaging typically includes conservation narratives of concern and responsibility for animals and natural systems and can thus be unexpectedly convergent with zoo mission goals.[30]

Identity Development Implications

An incredibly personal element of value, identity development can be instrumental in building connections between visitors, the zoo, and the

---

[29] Fraser and Sickler 2008b.   [30] Human 2005.

zoo's mission. The findings presented in this chapter are a starting point to explore the potential for strengthening the relationships between zoo mission and aspects of identity through which new pathways might emerge to address local conservation-related concerns with local solutions – rather than constantly conceptualizing conservation as a global goal that happens elsewhere, and is thus largely the responsibility of people and governments elsewhere. We have also begun to tease out the unique potential of zoos as regional recruitment centers for conservation agenda action, a legitimate opportunity *because* zoos are highly accessible and appealing spaces wherein diverse individuals and groups build relationships and negotiate meaning, identity, and values.

Deeper understandings of zoos as social and physical environments that impact identity and inform how various groups negotiate their concern for captive wild animals affords scholars, conservation psychologists, and zoo staff alike the opportunity to better understand the role animals may play in the development of social concern for the environment. Indeed, linking the psychological development of concern for animals to self-concept and social norms regarding wildlife makes it possible to consider how animal relationships inform both consciousness and decision-making regarding nature protection. In Chapter 10, we thus begin to look at how identity can also play a structural role in promoting action and explore zoos as activation entities.

Activation
Pro-environmental Behavior

Among the more interesting themes that have emerged from the plethora of studies on conservation values and impacts for zoo and aquarium visitors in the first twenty years of the twenty-first century has been the finding that personally observing living animals, whether the animals are within garden-like simulations of natural environments or "classic" hardscaped enclosures, tends to create for zoogoers a valued sense of connection to and continuity with the natural world. Chapter 9 introduced data suggesting that zoos offer valuable opportunities for forming, deepening, and contesting identity – the sense of who each of us is with respect to our own social networks; our place; other groups or groupings of people; the nonhuman animal world with all its complexities, commonalities, and categories; and the personal responsi-bilities we feel to "others." Indeed, John Fraser and colleagues have shown that zoogoers' self-directed experiences on site consistently intersect with the psychological dynamics of identity work, the personal process of under-standing oneself and one's behavior patterns.[1]

This may be why tests administered before and after zoo visits reveal little impact on conceptual understanding of conservation issues or reten-tion of natural history facts,[2] yet zoogoers who encounter poems on display in zoo settings have been shown to develop new ways of thinking about animals as "others" that shift their personal perceptions regarding who is (and who should be) responsible for active practices of animal care and conservation.[3] As sites in which social interactions are centered around animals, we showed in Chapter 9 that zoos have unique opportunities to enable the development of an environmental identity that reflects concern for animals. Importantly, though, even those who care strongly about animals and/or nature often fail to engage in pro-environmental behavior, a pattern known as the value-action gap.[4]

---

[1] Bruni et al. 2008.   [2] Moss et al. 2015.   [3] Fraser et al. 2008; Preston 2013.
[4] Bloodhart et al. 2013.

Returning to the idea of the zoo as a museum, we thus introduce a new layer of ethical narrative and analysis around the value and utility of captive wildlife: the process of activation that provokes an individual to shift from the related domains of ethical consideration and identity work into action mode. While direct, in situ conservation endeavors funded by zoos with zoo personnel support can have mission-consistent conservation outcomes and impacts, those actions are often far afield from the zoo's work holding, breeding, and displaying wildlife. On site, ethnographic researcher Irus Braverman has shown that zoos instruct and attempt to motivate visitors to care about (and for) far-away animals. She suggests that zoos do this by carefully associating zoo animals and zoo conditions with the zoo animals' non-captive counterparts and the condition of less-managed landscapes – and by simultaneously associating animal care with action.[5] Here, we thus explore at the potential of zoos as activation entities able to shift zoogoers from the process of individual identity work related to environmental conservation values to the active and ongoing pursuit of a conservation agenda.

Conservation Values and Identities, Precursors to Conservation Activism?

Since the first Earth Day ushered in the modern environmental movement in 1970, environmental organizations and activists alike have focused significant energy on changing human behavior. Unfortunately, while educational programs preparing "the next generation" for a lifetime of environmental stewardship have been evolving for two full generations, biodiversity loss and environmental degradation continue. It is thus important that we pause to reflect on what is needed to *activate* a conservationist, a line of thinking grounded in the content of earlier chapters. How does an individual shift from comparatively vague thoughts and partial understandings to moral conviction that conservation is a good idea – and then to consistent conservation mindset and action patterns?

Developmental precursors to moral concern and related actions of care have long been understood to include appreciation of fairness and welfare, sensitivity to standards, feeling with others, and the ability to overcome egocentrism and moral selfishness in situations unassociated with significant stress.[6] Environmental ethics researchers have found that individuals with a *strongly developed, internalized conservation ethic* enact this perceived

---

[5] Braverman 2013.   [6] Packer et al. 1985.

virtue through habits of action informed by conservation goals. Individuals with an *emerging conservation ethic*, on the other hand, will express that comparatively less-clarified virtue through increasingly routinized small and/or large decisions perceived to advance conservation goals. So, while anyone with some degree of conservation ethic will demonstrate a degree of perceptible action to advance conservation goals, conservation is a complex undertaking and will inevitably reflect a range of visions, priorities, and moral and operational frameworks greatly impacted by socialization.[7]

Like any social movement, the environmental movement encompasses numerous (often regionally distinct) discourses and includes both highly committed, highly engaged activists who seek to reshape the ideological and normative practice landscape and recognize the movement as a core aspect of their identity, and comparatively less engaged supporters sympathetic to the movement. As such, while the social-psychological implications of conservation ethics can be collectively harnessed, environmental activism is inevitably valued and expressed in myriad ways connected to each (moral) actor's circumstances, worldview, and willingness to pursue different actions. That said, the shared values that unify conservation movement actors relate to beliefs that: (1) human action often impacts the biophysical environment in ways that harm beings, things, and conditions people care about and value, and (2) action must be taken to mitigate or eliminate such harm.[8]

At the turn of the millennium two decades ago, researchers in the United States were suggesting that support for the environmental movement tends to develop as a linear progression.[9] The linear lens is supported by two late twentieth century theories of social change: the theory of planned behavior (TPB) and the value-belief-norm (VBN) theory of action in support of social movements.[10] In both cases, the focus is the individual's beliefs, understanding of various actions they can take, and sense that the action they take will result in the desired action outcome.

The theory of planned behavior holds that the pathway to activation involves (1) promoting what to do, why it's important, and that it can be

---

[7] See Clayton and Meyers 2009 for an in-depth analysis of theory and data related to psychological moral functioning.

[8] See Stern et al. 1999 as a starting point to contextualize the logic that grounds this conceptualization of social movements, and the environmental movement specifically.

[9] Stern et al. 1999.

[10] See Ajzen 1991 regarding the theory of planned behavior, and Stern et al. 1999 for a primer on value-belief-norm theory.

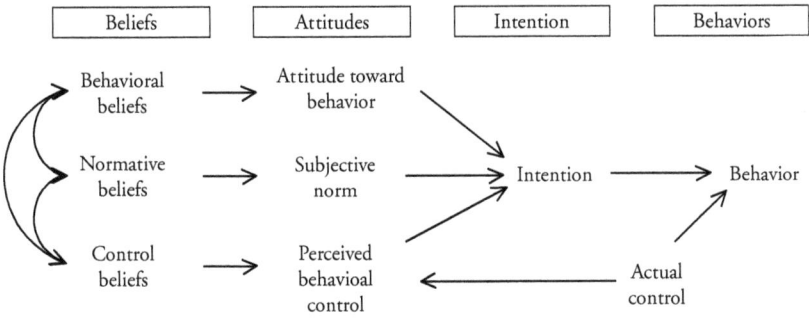

Figure 10.1   Theory of planned behavior

achieved by the individual; and (2) helping the individual overcome perceived barriers that situate the solution as outside the individual's control. TPB is a powerful tool for assessing and predicting single actions that produce individual-level reduction of environmental harms.[11] More than adequately proven, TPB has been cited more than 75,000 times in peer-reviewed journals in the thirty years since social psychologist Icek Ajzen first published on the topic.

While TPB can predict any type of behavior, from health directive compliance to seatbelt use, the value-belief-norm theory specifically highlights the roles of personal identity and the individual's working understandings of system conditions in their behavior decisions. Situating behavioral change as rooted in the person's sense of who they are and the values they hold, VBN theory suggests that individuals draw on their sense of self to make decisions that align their identity with their value priorities, and recognizes that activism tends to start with small steps in support of a movement.[12] VBN models do not appear to have as much predictive strength as TPB models, but both recognize that people rarely exist in a behavioral vacuum – context matters.

Pushing beyond movement sympathy, activated individuals begin to enact a willingness to take some action and bear some cost/s to advance movement goals, without necessarily committing to ongoing, highly engaged activism. Researchers have focused on empirically distinct types of individual environmental action pathway that signal social movement support. Low-commitment active citizenship might be enacted through

---

[11] Oreg and Katz-Gerro 2006.   [12] Stern et al. 1999.

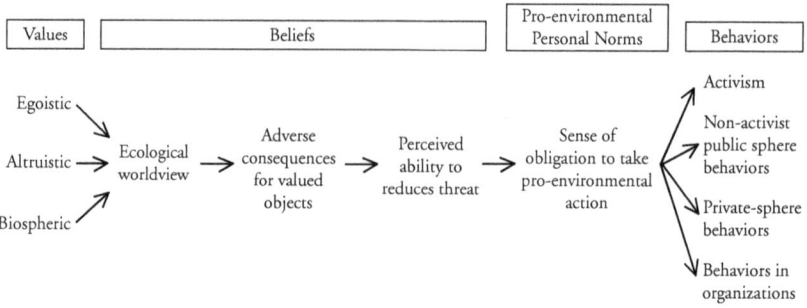

Figure 10.2   Value-belief-norm theory

written petitions to political officials; monetary donations or small amounts of time gifted to support movement organizations; and self-directed learning about the movement or movement goals. Personal support for and acceptance of policies perceived to support movement goals often require some degree of self-interest trade-off, typically in the form of inconvenience, higher prices, higher taxes, or behavior regulations (proscribed recycling, for example, or a cap on water consumption). Personal-sphere behaviors, such as committing to short showers or landscaping with native plants, often signal citizen perspectives and consumer preferences, and can be initiated and play out in private or in public.

Switching off a light bulb, for example, is a personal, private-sphere behavior choice, whereas personal public-sphere behaviors such as placing recycling containers at the curb in front of your residence for pickup are actions observable by others. Scientist Paul C. Stern, one of the creators of VBN theory, has noted that anticipated social dynamics associated with various types of behavior and behavior fora often inform the action patterns an individual chooses.[13]

In practice, it is exceedingly rare for individuals to become activated or engage in activism (behavior outside social norms) without social support. Conservation psychology research has shown that individuals become increasingly willing to actively pursue a conservation agenda as aspects of their emerging or established environmental identity are accepted, affirmed, and nurtured. Recent studies also suggest that individuals may become *de*motivated if they interpret environmental action pathways as

---

[13] Stern and Dietz 1994.

personal sacrifices and think other individuals they perceive as similar are not making comparable sacrifices.[14]

On the flip side, though, pleas for political action can inhibit people who feel that political activism is ineffective, a pattern likely to be identified as a lack-of-control belief obstacle in a TPB behavior prediction model. Lack-of-control belief obstacles can be particularly formidable barriers to widespread action on complex conservation matters such as climate change, and we note with concern that mitigation has frequently and problematically been framed (and failed to move the needle) as large-scale political debate, rather than scientific reality that can be worked on and disrupted at local and regional levels if solutions are valued.

To achieve social change, activists can be willing to push their public and private personal-sphere behaviors in multiple directions they have access to, including, in some cases, within organizations where social innovation is nurtured through common agreement with others who accept them as part of a group. A range of studies focused on community-based social action have shown that people wish to behave in ways others will likely identify as in-group behaviors. Knowing one's neighbors are engaging in more conservation behaviors, for example, often results in greater awareness and efforts to adapt one's own behaviors in various ways.[15]

Zoos As Activation Entities

While theaters, sports venues, places of worship, public parks, and theme parks are other examples of social experience spaces that attract people during their leisure time, museums are unique as venues where one's civic responsibilities to the community, to other beings, and/or to the biosphere are explicit and integrated topics of discourse. As institutions operating in the public domain as social fora for explorations of the meaning of community-level civility and the community's life-giving choices related to the environment, zoos are museums that have community-granted authority to tell stories about and present artifacts and evidence of what it means to be a civilization.

Zoos, which once offered visitors chances to view animals out of their natural context or within a presentation of idealized nature, now seek to create a sense of animals as important parts of specific places and represent biosphere realities through lenses of human connection and responsibility.

[14] Bloodhart et al. 2013.   [15] Schultz and Tabanico 2007.

This context provides visitors with opportunities to reflect on their connections to and involvement with larger patterns and systems through free-choice exploration of various aspects of those dynamics. Rather than focus on the level of the visitor or situate a zoo visit as a single moment in time, we suggest that zoos offer social value as places where ethics are explored within and across groups, in distinct settings that support the navigation of multiple ethical codes, meanings, and potential responses. People visit zoos for and with people, and, while there, tend to discover what norms and decisions appear to be consistent with the social acceptance they seek.

Animals are a bridge to caring about the natural world, according to many theorists, not only because nature is a source of socially constructed symbolism from which people derive meaning, but because nature and animals are enduringly compelling to humans of all ages, eras, and cultures. Dynamic and highly responsive, live animals appear to have specific appeal as the only nonhuman entities that allow humans to have direct experiences with a feeling "other." Encountering living animals also nudges people to situate themselves in relation to what they can observe about that animal's life in the world. Integral to sense of self during childhood, such experiences often lead to care-focused outward actions.[16]

Given that zoos seek to encourage protection of nature based on zoogoers' experiences with animals on site, we seek to establish the degree to which zoos appear to help zoogoers overcome social norms inhibiting them from including animals in their scope of justice and acting on their concerns about animal well-being. In particular, we explore the degree to which zoos can advance injunctive norms that acting to protect wildlife is desirable and reasonable, given most zoogoers' beliefs and values, and that action to work toward that moral social norm is within the control and capacity of individual zoogoers.

Zoos could be ideally suited to appeal to their visitors based on the value-belief-norm theory of action *if* data suggest visitors' entry values align with the conservation asks made of them by the zoo, particularly for visitors who already strongly affiliate with animals, as such visitors whose values have logical intersections with a conservation agenda that can likely be activated through zoo visit experiences into a willingness to establish action patterns that more closely honor those values. WZAM data show that zoos are valued as settings/experiences that inspire moral reflection and moral development among visitors of all ages, and the great majority of zoo visitors have a high level of comfort with zoos as sites that provoke public audiences to connect human attitudes and behaviors to the well-

---

[16] See, for example, Stern 1985 and Melson 2001.

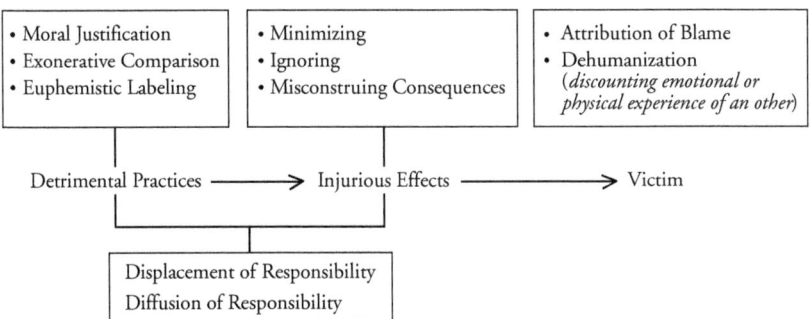

Figure 10.3 Psychosocial mechanisms that allow for selective disengagement from detrimental conduct in the moral control process

being of the environment and the animals therein. Furthermore, adult visitors specifically reported valuing zoos for providing them with explicit messaging and guidance about behavior changes that will help them become better stewards of the environment. While the data suggest that zoo visitors are not necessarily seeking to be formally educated about the scientific foundations of environmental issues or animal-specific phenomena, most do wish to become more environmentally aware and be given "rules of the road" that they can use to become more environmentally responsible in their own lives.[17]

Psychologist Albert Bandura's exploration of peace psychology sheds further light on how zoos and other conservation actors might nudge public audiences to action by highlighting some of the tendencies and behaviors that exclude or preclude consideration of an animal as being within one's scope of justice. As one example, Bandura points out that when a term or phrase justifies harm or exclusion, misconstrues impacts, or dehumanizes an individual or group of people, language can become a powerful tool for moral exclusion.[18] We see this when food animal and domestic livestock terms such as pig, cow, dog, and goat are invoked to describe humans as being unworthy of moral consideration.[19]

Social psychologist Susan Opotow suggests that developing a sense of empathy toward an animal is one path toward inclusion of animals in people's scope of justice, even if prevailing social trends might lead to that animal or type of animal being excluded.[20] Such deliberations are, in fact,

[17] Fraser and Sickler 2008b. [18] Bandura 2012. [19] Beirne and South 2012.
[20] Opotow 2011a, 2011b.

evident in the language we hear when people visit a zoo and in the growing body of data regarding zoogoing and zoogoers' experiences.

WZAM research conducted in collaboration with Colorado State University shows that zoogoers are much more likely to include wildlife in their scope of justice than the general population. Using scales developed by Michael Manfredo and colleagues, a series of studies assessing value orientations toward wildlife between 2008 and 2019 have shown that zoo visitors in the United States consistently held more mutualistic value orientations than individuals in the broader community population, irrespective of respondents' degree of urbanity or which region of the country they lived in. Zoo visitors in the WZAM2 sample, for example, valued wildlife for its existence (intrinsic) value more than the general public, and reported being more willing to support protected status for wildlife and more likely to support policies to protect wildlife habitats even if that protection would involve paying taxes or fees.[21] We also now know that people who visit zoos and aquariums are more politically engaged, more environmentally concerned, and more motivated to take action than non-visitors, even if their knowledge about environmental solutions does not exceed that of their neighbors.[22]

From an activation perspective, then, zoo visitors overall are "users" who arrive prepared to take a step forward.[23] Over twenty years of WZAM data have demonstrated that most visitors arrive at zoos and aquariums with foundational knowledge about basic biological concepts but appreciate and make meaning from zoo experiences that engage their senses and emotions and provide new points of reference. Furthermore, most zoo visitors report seeking specific information about how they can become more personally involved in conservation advocacy and make a bigger impact, and describe meeting their on-site learning needs through strategies and forms *other than* formal science knowledge gain.[24] When zoo and aquarium staff respond to surveys about their dialogues with visitors, though, most report focusing on animal science and natural history. While visitors report finding such information interesting, John Fraser notes that early analysis

[21] Saunders et al. 2005. We note that these findings were consistent with prior research showing that personal perceptions about the "justice" of proposed environmental policies are better predictors of support than the extent to which individuals believe a policy will impact them personally; see, for example, Kals 1996.

[22] Swim et al. 2017.

[23] David Carr 2011 positions those who visit museums of any type as users who arrive at the institution ready to advance their own agendas, suggesting that museums meet them midway on their personal learning journeys.

[24] Fraser and Sickler 2008b.

of the WZAM3 data set suggests that most visitors are more interested in getting closer to solutions they feel they could achieve. This suggests that there is great potential for integrating into existing programming tangible conservation action suggestions and resources that zoogoers can walk away with.

Despite this apparent disconnect, which points to important opportunities to adopt and fine-tune more effective models of visitor engagement, mission alignment, or at least the foundation for mission alignment, does appear to occur through zoo visit experiences. WZAM data suggest that most visitors are leaving the zoo with a stronger sense that they personally have a role to play in mitigating environmental degradation and feel they can do many things to advance a conservation agenda. In fact, most zoo visitors report seeking specific information about how they can become more involved in conservation advocacy and make a bigger impact.[25]

Implications

In order to activate and subsequently support the ongoing activism of individuals whose priorities, circumstances, and willingness to pursue different actions vary widely, zoo leaders, staff, and volunteers must envision and situate themselves and their institutions as actors within a much larger conservation movement. Just as any complex problem tends to benefit from multiple treatment strategies, actors seeking the widespread adoption of a conservation ethic, including zoo staff, need to recognize the importance of connecting public audiences with multiple, complementary action opportunities to achieve conservation impacts that notably reduce the scale and scope of the current crisis.

It is thus encouraging that zoogoers are seeking out resources and tools that will help them continue learning about and engaging in conservation action opportunities *after they leave* the zoo. In addition to pointing visitors toward websites, films, television programs, blogs, books, and social media resources that might extend their knowledge and/or further pique their curiosity, zoos should be connecting visitors with conservation action and advocacy groups that are both local and further afield.

Historically, zoos have sought to activate zoogoers using suggestions of individual-level curtailment, substitution, or technical innovation grounded in morality and a (hopefully) convincing case for cause and effect. The thinking is that *curtailment activation* occurs when the

[25] Fraser and Sickler 2008b.

individual is nudged to elect to limit/reduce or eliminate a behavior understood to be problematically harmful. *Substitution activation* pathways typically focus on choices to consume products/inputs that are comparatively benign (e.g. lower-energy LED lighting or high-efficiency appliances) in order to mitigate problematic harm. *Technical innovation activation* is understood to occur when an individual is successfully nudged to replace a more harmful solution with a less harmful solution to achieve the same outcomes (e.g. switching from regularly driving to work to taking public transportation).

Data amassed over the past twenty years clearly show that the users of zoos and aquariums seek visions of near-future sustainability pathways and are interested in being acknowledged as partners in conservation. We note, though, that in too many zoos visitors continue to be conceptualized as individual targets the zoo wants to change through mission-related messaging primarily limited to "don't" suggestions aimed at individuals and focused on denial.

While the psychology of denial as a pleasurable experience can work effectively for a limited population, the average individual does not respond well to denial appeals.[26] Though the SUVs in the parking lot and demand for imported plastic souvenirs may suggest to the engaged conservationists who work in zoos and aquariums that visitors should be exhorted to change their ways, such appeals can map to the temperance movements of religious history. Whether the focus has been alcohol or abstinence, appeals of denial have shown limited success over the decades as overlays for leisure choices. Accurate but narrow on-site messaging about the overwhelming harms of fossil fuel reliance, for example, can inadvertently create feelings of disempowerment and/or guilt, especially if the context is further complicated because exhibits in the zoo were funded by regional businesses tied to fossil fuel consumption, and those businesses employ large numbers of zoo visitors.

Substitution strategies salient to the zoo experience may align more closely to a user's needs and desires. Substitutions provide opportunities to negotiate within a group with respect to weighing and comparing values and goals. Environmental psychologists have demonstrated that people are more likely to fear loss than unwilling to try something new, and while admonishing antienvironmental behaviors can be effective interpersonal behavior for eliciting desired actions, subtle encouragement to alter behavior to better align with a pro-environmental agenda can be even more

---

[26] Herziger et al. 2020.

effective. This is particularly the case if the target audience attributes the encouraged actions to their own (desirable) environmental attitude. Recent research suggests that nudging people toward pro-conservation attitude and identity affirmation is an effective mechanism for creating behavioral consistency, and behavioral consistency has been shown to be more important for changing social norms and behaviors than praise.[27]

This type of nudging strategy was explored when a composting rest-room and grey water garden exhibit was established at the Bronx River Gate of the Bronx Zoo in 2006. The facility conserves significantly more water and energy than a typical public restroom and was envisioned as a space for sharing conservation messaging with visitors of all ages regarding sustainable practices that benefit humans and animals alike. Signs created by the design team highlight fun facts about human waste, how human waste affects the environment, how composting toilets work, the sustainability value of composting toilets, and simple actions to conserve water and energy on a regular basis.

A survey was administered during discrete periods in 2006 and 2007 to identify visitors' impressions of the eco-restroom, whether learning occurred through experiences in that space, and whether visitors anticipated incorporating focal concepts into their lives. Among the 172 survey respondents, 86 percent reported that they read the signs and over three-quarters described learning at least one new idea or concept. Nearly all respondents left with a positive impression of the eco-restroom and agreed that it creates an entertaining, comfortable, informative, and empowering environment. Furthermore, and in keeping with WZAM survey findings about visitors' positive disposition toward (and expectation of) conservation information and messaging at the zoo, 88 percent of respondents reported that this exhibit experience made them feel good, appreciative, and hopeful. Importantly, over half of the zoogoers surveyed anticipated incorporating new conservation concepts into their lives, suggesting that this experimental merging of sustainability-focused exhibit design and scatological humor had activation impacts. Nearly all visitors who completed the survey noted that the eco-restroom improved their overall impression of the Bronx Zoo, and it continues to be a popular media focus and a voter favorite in online tourism voting competitions.

After years of exhorting zoo visitors to commit to (at least) one conservation-focused action, a few zoos are finding success as exemplar community leadership sites. Two Melbourne Zoo projects are particularly

---

[27] Swim and Bloodhart 2013.

compelling and rare examples of efforts to activate an interested community. The first, a substitution effort appealing to the moral logic and environmental impact of replacing heavily processed toilet paper with unbleached toilet paper, is as an example of long-term community-based social marketing that focuses both on individual behaviors and injunctive social norms.[28] The second is a program that presents the threat of wide-scale palm oil production to orangutan habitats while seeking to leverage zoogoers' affinity for nature to promote less harmful consumer behaviors.[29] The success of these projects suggests that when given compelling tools and information, affinity audiences can be activated during a zoo visit to engage in individual-level conservation actions after they leave the zoo.

Providing visitors with immediate opportunities to participate in pro-conservation behavior has been shown to successfully convert behavior intentions into action.[30] The individual, though, is not the only relevant locus of control for activism. Together, people can help each other see and articulate any human behavior as a pattern that can be disrupted. Curtailment, substitution, and divestiture are all examples of conservation action pathways that can be effectively (and simultaneously) pursued through collective mechanisms at the organizational, community, state, regional, and international levels. We therefore turn in Chapter 11 to the mobilization potential of zoos, exploring the extent to which these trusted, place-based cultural institutions appear to have the power and capacity to connect zoo users with collective-level conservation identity and engagement opportunities.

---

[28] The Melbourne Zoo has documented a long-term effort to change the toilet paper choices of members and visitors through their Wipe for Wildlife program, a social commitment platform for Australians declaring willingness to purchase 100 percent recycled unbleached toilet paper: www .zoo.org.au/wipe.

[29] Pearson et al. 2014.   [30] Gwynne 2007; Powell and Ham 2008.

CHAPTER 11

Impact
Collective Conservation Action

When Knology, the think tank with which both authors are affiliated, set out to help zoo and aquarium interpreters develop skills for talking about climate change, we knew the topic was emotionally loaded. Citizens of the United States had endured decades of vicious attacks against and denial of the abundant evidence that human production of carbon has caused irreparable harm to the biosphere. Yet, zoos were calling on their frontline interpreters to present the story of climate change because all reports indicated that there was a knowledge gap in this critical environmental reality. Working with colleagues at the New England Aquarium, Frameworks Institute, and Woods Hole Oceanographic Institution, Knology (then New Knowledge Organization Ltd.) helped launch the National Network for Ocean and Climate Change Interpretation (NNOCCI), an aspirational strategy for depoliticizing climate science and helping interpreters simplify and explain it in ways that helped diverse audiences engage.

At the outset, in 2010, we knew that those joining the NNOCCI training program felt that climate change was a difficult issue to raise, and many of our recruits admitted that they actively avoided talking about it, primarily because they lacked confidence in their knowledge of the topic and were afraid they would be open to personal attacks by climate deniers.[1] The depth of participants' emotional experiences fell open for us when one participant started to cry. At first, our training team was concerned that negative emotions were overwhelming the participants, but the reverse turned out to be true.

In tears, our mid-career interpreter looked around, at a room of about twenty-six people from zoos across the country, and shared that she was realizing for the first time that she and her peers had the reach, skills, and capacity to change public narratives about climate change. The tears were

---

[1] Swim and Fraser 2013, 2014; Swim et al. 2014.

recognition that what had seemed like a hopeless obstacle was suddenly within the grasp of the people in the room. This sense of hope, grounded in the realization that zoo educators could be the linchpin that finally releases public action across the nation, was a feeling she had never experienced in her career. It turned out she was not alone, and we have reason to believe her hope is well founded.

In recent decades, museum research has started to revisit philosophies, first articulated in the early twentieth century, that museums create opportunities for community advancement.[2] Scholars have examined, for example, how social context impacts individual learning in museums to explore how zoos and other cultural institutions are connected to the cultural replication of ideas within society at large.[3]

Psychology data point to identity and collective action as potentially convergent experiential pathways that are highly relevant to zoo goals. Water, wildlife, and forests are examples of valuable natural resources that cannot be effectively managed without collective action. Here we focus in a fresh way on the zoo movement itself, exploring how complementary and collective work across the sector can produce society-level change – drawing attention to how these processes and impacts are a far departure from the idea of inspiring an individual, through a single zoo visit, to individual-level action that might hopefully add up to something meaningful, at some point. While scientific knowledge transfer can be part of this engagement process, we move past the vision of zoos as sites of individuals' science learning, demonstrating that zoos have the capacity to impact society in broad, coordinated, and tangible ways through coordinated education programs and messaging that builds on emotional and relational connections to spark collective conservation action that may move the needle on conservation outcomes.

This chapter outlines evidence that staff working at a distributed network of zoos, aquariums, and nature centers can deliver coordinated conservation education that results in changed environmental ethics at the community, regional, and national levels. We close by introducing community of practice programming designed to engage and sustain concerned community members in deliberation, dialogue, and collaboration around priority conservation concerns – a social radiation approach that captures the powerful potential of tiny publics around the world.

---

[2] Dewey 1963 and Dana 1999 are among the examples; Hein 2004 provides a summary of the progressive philosophy of museum education.
[3] See, for example, Wilson 1992 and Archibald 2004.

Identity and Society

As we have shown in previous chapters, experiences in zoos have been examined for their contribution to zoogoers' individual development of identity and social capital. These studies, however, were not intended to measure or in any way account for the cultural value of zoos. Existing identity research frameworks support the cultural popularity and psychological value of zoos at an individual level, but individual identity development does not speak to the larger social value of zoos as cultural catalysts.

As we noted earlier, zoos are perceived by the general public as primarily social experiences.[4] The social nature of these experiences suggest that the individual as a unit of measure may misplace the deeper value of these destinations in society. The need to belong in groups is a fundamental psychological motivator.[5] Social identity, group identity, and collective identity are broad areas of research that seek to link theories of self to the structure and function of social groups, and although the language used in disciplinary-specific studies tends to be siloed and appears contradictory, there are many commonalities between the significant studies in sociology and psychology.[6] As a multidisciplinary framework, conservation psychology can unite disparate areas of inquiry toward the goal of understanding how care for the natural world is developed within society.[7]

A core focus of conservation psychologists has been the identification of theoretical models that both accurately situate and can be used to support the development of social and cultural norms that advance a conservation agenda. Psychologists have long pointed to the role and value of group activity in fostering individual motivations to protect nature,[8] and collective identity seems to have emotional significance and value because categorical membership involves connectedness of fate with others in the group as well as to oneself.[9] While social role and consequential actions have been situated in the past as behavioral implications of the identity

---

[4] Pekarik et al. 1999.   [5] Baumeister and Leary 1995.

[6] Ashmore et al. 2004 proposed a common nomenclature drawn from across the disciplines to establish a common measurement strategy for identifying individual-level elements that reveal collective identity in order to predict outcomes. The framework these scholars established features integrative categorization for collective identity research divided into self-categorization, evaluation, importance, attachment and sense of independence, social embeddedness, behavioral involvement, and content and meaning. These categories, the authors suggest, can be analyzed to determine how situation and context may create an interdependent set of study variables influencing identity development and outcomes.

[7] Saunders and Myers 2003.

[8] See, as starting points, the edited volumes by Roszak et al. 1995 and Clayton and Opotow 2003.

[9] Tajfel 1982; Tajfel and Turner 1986.

schemas within which a person operates,[10] an increasing number of social psychologists and sociologists now feel that group-based and collective identities (identities important to self-concept but commonly agreed to be shared with others) show the most promise for understanding motivations for social change and political action.[11]

Metatheoretical analysis of social identity research has outlined the value of studying small groups in specific situations to demonstrate how people can operate at more than one categorical level, assimilating multiple collective identities into one subject.[12] This theory suggests that group process analysis can tease apart the value of groups in crafting a variety of social conceptions of the self. These processes are dynamic, adaptive, and complex systems that increase in complexity based on the experiences and situations encountered by the group.[13]

An identity construct outlined in Chapter 9, environmental identity has been described as situating the self within the natural world, evident through the degree to which care and concern normally accorded to other people is extended to nonhuman life.[14] Systemically acting as if these elements of nature are part of the self has been defined as the exemplification of personal ecological identity,[15] and has been shown to be a measurable element of self-concept.[16] The extension of self to include environment and nonhumans in social and interpersonal relations may represent aspects of a collective identity that begin to accept oneness with nature. In such contexts, environmental identity may not be something achieved as an individual goal, but the result of a group process and series of contextual experiences that eventually form a bond to nature as part of the normal social process of building a sense of belonging with others, including other animals.

When people reassess self-concepts and identity frameworks related to expressing concern toward or about animals and share these types of experiences and thoughts with others, they often reveal a process of recognizing and recalibrating group values. Groups may also include themselves in a larger group or category based on recognizable commonalities of sentience or demonstrations of cognitive abilities.

Collective Identity

Collective identity is a sociological concept situating the degree to which an individual's perceived self-worth is inextricably tied to a group they

---

[10] Stryker and Serpe 1994.   [11] Brewer 2001; Fine and Harrington 2004.
[12] Abrams and Hogg 2004.   [13] McGrath et al. 2000.   [14] Thomashow 1995.
[15] Zavetoski 2003.   [16] Clayton 2003.

value or feel emotionally attached to. Research suggests that social group membership influences the likelihood that an individual will tend to seek to behave as a typical group member, often adopting the shared values and beliefs of a group after they have begun showing behavioral characteristics typical of the group.

Groups visiting the zoo are typically comprised of an established in-group with preexisting coherence, though a zoo visit often provokes further exploration of the dimensions of the collective identities salient to the group. The applied study of collective identity theory in the context of zoo experiences may reveal identity formation pathways that remain concealed when the focus is on individuals, rather than a level of socio-logical importance that points to how social public processes are involved in the development of cultural values. Specifically, understanding how families and other social groups at the zoo explore environmental values that are part of their collective identities may offer insight into the value of zoo animals as a contextual influence for group negotiations of concern for animal and environmental well-being.

Collective identity has emotional significance and value because collec-tive identity ascribes categorical memberships and connectedness of fate with others in the group to the self. According to sociologists, memorable experiences become *cultural* through discourse and repeated references, and narratives or objects linked to specific times and places facilitate people's recall of fun or enjoyable moments, fostering and affirming collective identity.[17] More nuanced understandings of the value of zoos in constructing social environmental narratives will better position zoos to be intentional about linking the zoos as sites of learning and emotional connection to the public framing of social behaviors that advance a conservation agenda.

Environmental Identity, Developed Collectively at the Zoo

Public support is crucial to social movements seeking to disrupt cultural inertia and the interests of powerful actors. In recent decades, museums have been re-characterized as the new civic square – spaces that bring communities together to develop shared civic concepts.[18] Zoos, further-more, have been identified as a unique museum type that appeals to adult groups as a place to pursue social, in-group objectives rather than the more

---

[17] Fine and Corte 2017.   [18] Karp et al. 1992; Archibald 2004.

personal cognitive and introspective goals adults associate with other museum types.[19]

We note that while publicly funded zoos have been very reasonably identified as institutions that replicate the social values toward nature of the culturally dominant elite,[20] the social context of zoogoing does appear to consistently incorporate nonhuman animals and natural place concepts as part of a given collective identity. Understanding how zoo-visiting groups negotiate dis/agreement regarding their relationships to animals they see, how they accord these animals' rights, and the degree to which they seek affiliation with animals and with each other can help shed light on how environmental values are evolving (or might evolve) at the community and society levels.

Family value negotiations and other collective identity explorations within a social cohort are evident in personal actions and conversation performed in public. Family values transfer is a notably significant aspect of cultural identity, which is why so many researchers have explored the patterns and potential implications of adult visitors' use of museum experiences to bond with and inculcate family values in their children.[21] Live animal displays at the zoo provide both novel and predictable stimuli for family and other groups to explore the context of their in-group relationship, often with respect to frameworks and boundaries for collective identities around concern about, care for, and/or inclusion of animals.

To understand how collective environmental identity research can be investigated at the zoo, it is useful to imagine an exemplary social condition characterized by such explorations. While many groups subtly reinforce dominant cultural values that can be difficult for a researcher to extract, culturally distinct groups sometimes reveal more obvious examples of how experiences with captive animals are negotiated. Imagine a Hindu couple from Mumbai who settled in the United States, where their children were born. This hypothetical couple can be imagined visiting a New York zoo in early spring with vacationing relatives from India. The context could provoke collective identity establishment, contestation, or reinforcement that would be directly tied to experiences of being at the site together. It is easy to imagine robust in-group discussions about being vegetarian in a cafe serving burgers and chicken fingers, the cultural relevance of the Asian tiger exhibit, childhood memories of direct encounters with live animals, different views and experiences of finding and

---

[19] Morgan and Hodgkinson 1999; Pekarik et al. 1999.   [20] Wilson 1992.
[21] Kidd and Kidd 1996; DeVault 2000.

building community, and what display techniques and zoogoer behaviors at zoos in different parts of the world reveal about the values and norms of local dominant culture.

Zoo organizing principles and displays offer a fascinating context for uncovering specific content and meaning ascribed to individual animals, the culture of nature, and how "wild" is conceptualized within a group. Wild nature can serve as a catalyst for restoring family kinship, bonding, group cohesion, and social recognition,[22] and it appears that the collective identities developed or reinforced by social, nature-based experiences, whether in the wild or in the constructed nature setting of a zoo, can spark collective identity development and potential action outcomes as long as the animals therein are living beings. How families and other types of groups frame their concern for animals also highlights how cultural norms come to be replicated or contested (if a group becomes motivated toward social action to change specific values).

Studies of the conditions that provoke such negotiations have the potential to reveal self-attributed characteristics regarding concern for nature, ideologies to which the group subscribes, and shared narratives regarding the value of wild and captive animals. Sociologist and animal ethics specialist Linda Kalof demonstrated through a qualitative study of college undergraduates that social discourses regarding animals are multi-layered and not necessarily parallel. She suggests that some causal roots for these discourses might be related to social, family, or ethnic experiences where values were inculcated during childhood and adolescent development. Her theory suggests that repeated family cultural and social experiences serve to frame social values toward animals and how a degree of concern is accorded to different types of animals in categorizing them as pests, food, useful, or beautiful.[23]

Moral critique may even provoke the incorporation of natural entities such as nonhuman animals or natural places into the identity work that occurs during a zoo visit, extending existing collective identity dynamics to include environmental identity dimensions. Zoos housing animals in clearly poor physical and emotional health have often provoked families and other groups to develop shared agreement that manifests as outrage, motivating them to take actions such as public protest, petitioning and pressuring others in positions of influence, or advocating in the press on behalf of the animals. Observations by that same group at another zoo may

---

[22] Roggenbuck and Driver 2000.   [23] Kalof 2000.

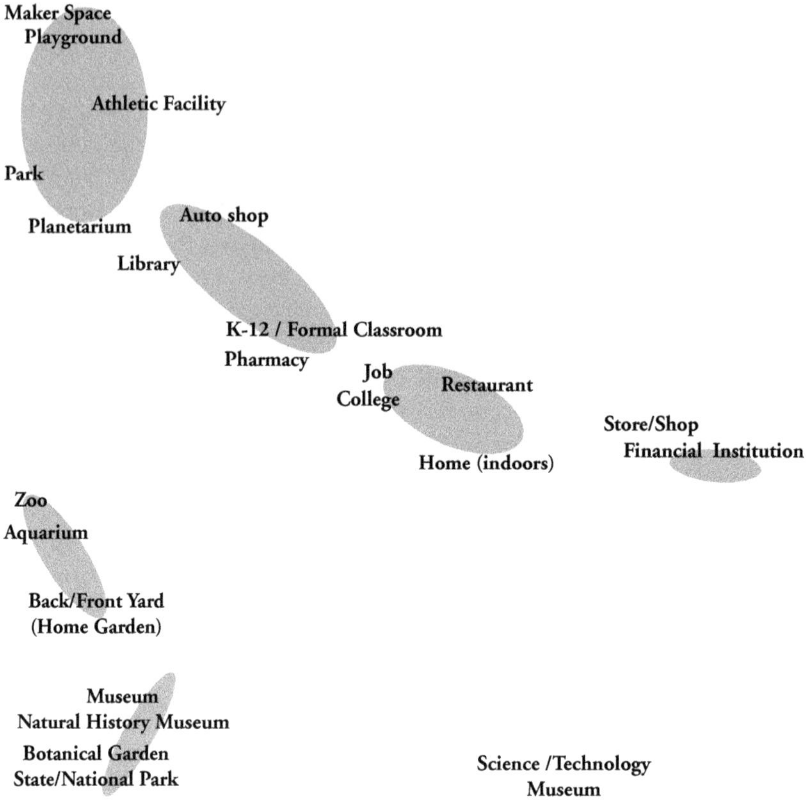

Figure 11.1   Cluster analysis output illustrating how research participants associate types of STEM learning sites

lead the same people to believe the animals are happy and content, resulting in a shared affinity with that zoo and defense of the zoo management against animal rights protestors. So how can the zoo serve as a stimulus for the social group to negotiate shared values regarding conservation and the development of a collective conservation identity and activism ethic?

Wildlife Value Orientations

The lore around zoo visitation is that zoos are demographically representative of their communities, although attendance demographics at many zoos tends to skew toward slightly higher incomes and zoo visits tend to be more often organized by female heads of household.[24] The first inkling that there may be a psychographic variation between zoo visitors and the general public emerged in the mid-2000s when John Fraser collaborated on two projects exploring the wildlife values of zoo visitors. The first set of studies applied a heuristic survey instrument developed by Ron Meyers to assess the issue of rights of nature across a variety of types of natural entities, from gorillas and apes to snakes, rocks, and insects.[25] The data showed that zoogoers were much more concerned about the rights of animals (and, indeed, *all* natural entities, including rocks) than the general public, and were uniformly more concerned about the conservation of nature than people in their neighborhoods who did not spend time in zoos. Interestingly, then, the attitudes of zoogoers are more closely aligned with those of anti-zoo animal activists than to the attitudes of the general public, with respect to concerns about the rights of animals. The research team then compared visitors to the New York Zoos and Aquarium with the greater New York City general population to determine whether the national Wildlife Values Orientation instrument[26] revealed more empathetic concern for wildlife among zoo visitors. In 2019, we were able to verify our unpublished results with a national study demonstrating that zoos across the United States have visitors who are significantly more likely to demonstrate altruistic value toward protecting wildlife and nature than non-visitors.[27]

Australian researchers Roy Ballantyne and Jan Packer were also working to advance the field-wide effort to understand the logical fit of conservation education as a prime role for zoos throughout the 2010s. Their studies confirmed that zoos and aquariums are already stereotyped as conservation organizations, with various nuances related to topical authority and a clear need to map current practice to a more highly engaged audience, and their findings have consistently suggested that zoos and aquariums are underperforming for the degree of conservation education desired by their visitors.[28]

There has been, in the past decade, a concerted effort to consider zoos and aquariums as part of a continuum of places associated with STEM (science, technology, engineering, and math) learning. While the work of wildlife conservation is most often associated with conservation biology,

[24] Findings of a multi-zoo study led by Joe E. Heimlich of COSI in Columbus, Ohio validated these generally accepted norms in 2020; data specifics will be detailed in a forthcoming publication.
[25] Meyers 2002; Saunders et al. 2005.   [26] Manfredo 2008.   [27] Fraser et al. 2020.
[28] Packer and Ballantyne 2010; Ballantyne and Packer 2011; 2016; Ballantyne et al. 2011; Packer and Ballantyne 2012; Ballantyne et al. 2018.

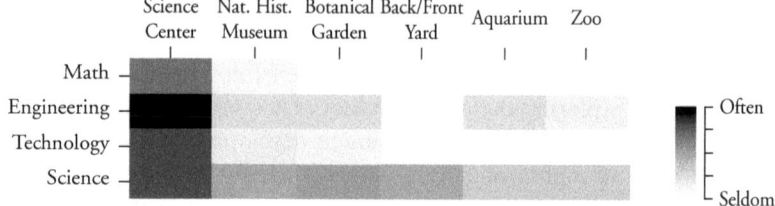

Figure 11.2   Public perception of the degree to which members of the public associate
frequency of encounters with STEM topics with different venue types

we have shown throughout this book that living collections create unique
opportunities for meaning-making that transcend the domain of science.
A study about institutional stereotypes, published in 2020 by Rupu
Gupta, John Voiklis, and colleagues, has shed additional light on zoos'
distinctness. The research team invited members of the public to reflect on
the nonschool contexts in which they encounter and learn about STEM
concepts and content, grouping these types of places according to which
seem most the same. According to the data, which was collected in the
United States, members of the public tend to see science and technology
centers as the sites most directly associated with STEM learning, and this
venue type was grouped on its own. Zoos and aquariums, on the other
hand, were clustered with the nature people can experience in their own
front and backyards (see Figure 11.1). This fascinating finding suggests
that many people may think about zoos and aquariums as spaces where
learning about and through STEM seems/feels accessible, casual, personal,
self-directed, observation-oriented, maybe even familiar. The same study
affirmed that public audiences expect zoos and aquariums to have
specific topical expertise that tends to go beyond wildlife to include the
technology and engineering topics that are part of wildlife management in
the zoo and are a platform for talking about how to protect nature (see
Figures 11.2–11.4).[29]

Potential of the Tiny Public

Individual and household actions, such as purchasing energy efficient
vehicles and appliances or weatherizing one's home, can be relatively easy
to commit to and coordinate, do not require new technologies, and can

---

[29] Rupanwita et al. 2020.

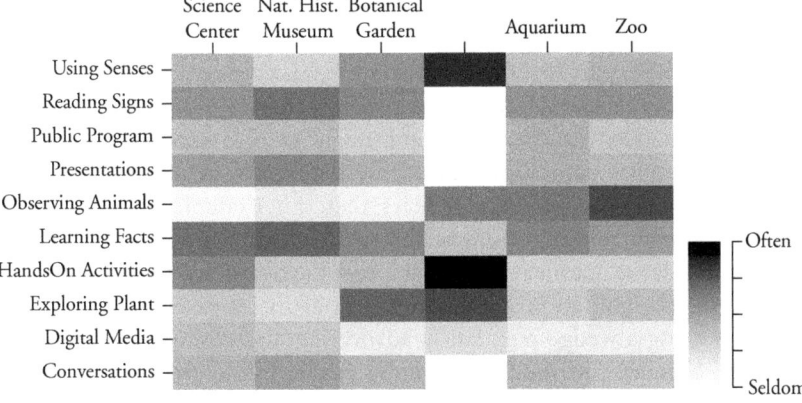

Figure 11.3   Degree to which members of the public associate frequency of encounter with a type of learning experience and place

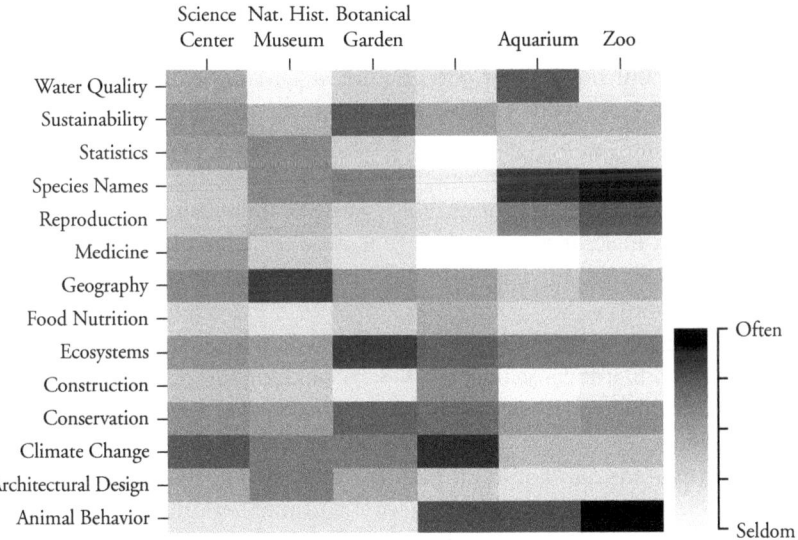

Figure 11.4   Degree to which members of the public associate frequency of encounter with a learning topic and place

funnel into significant change if overall demand for energy decreases. Unfortunately, though, some such actions can lead people to believe they are doing their part to stem environmental degradation but are too insignificant to make real difference.

Collective action, on the other hand, might involve working with others to solve complex community problems or engage in activism to shift political dialogue to promote policies that, for example, regulate heat-trapping gases or increase the availability of solar and wind energy generation. This latter pathway has long been described by thought leaders focused on the politics of the environment as the most effective field for action because a wedge population advocating and voting for policy and ideology changes can produce systemic changes that eventually become adopted as social norms.[30] Psychologists, however, caution that policies that cause perceived deprivation or are deemed to be illegitimate can result in ire, noncompliance, protest, and/or activism.[31]

Between these two extremes are coordinated efforts by groups of individuals. Sociologically, it has been shown that small, coherent groups can germinate social change.[32] Cultural theorist and social psychologist Gary Alan Fine and sociologist Brooke Harrington have shown that *tiny publics* can frame concerns in societal terms and motivate action toward social change, suggesting that small groups can be key vectors for social change and may present the greatest potential for advancing social change.

Neighborhoods, work groups, and nonprofit organizations have been particularly successful at advancing household-level changes (such as initiatives to weatherize low-income housing) and creating and supporting local solutions (such as community gardens, bike lanes, and farmer's markets) that make it easier and more natural for large numbers of people to take advantage of opportunities that reduce their environmental impact footprint.

The tradition of predicting environmental activism has tended to focus on an individual's consciousness that something of perceived value is threatened, a logical assumption emerging from the stories of exemplary activist leaders. Yet, small and incremental actions taken as part of a group are an important and common path for most pro-environmental behaviors. While studying volunteers at zoos[33] and undertaking a smaller study of activists who took up arms in anti-evolution protests at zoos and other science centers,[34] John Fraser found that the majority of public actions

[30] Spitzer and Fraser 2020.   [31] Moar 2016.   [32] Fine and Harrington 2004.
[33] Fraser 2009b.   [34] Fraser 2006.

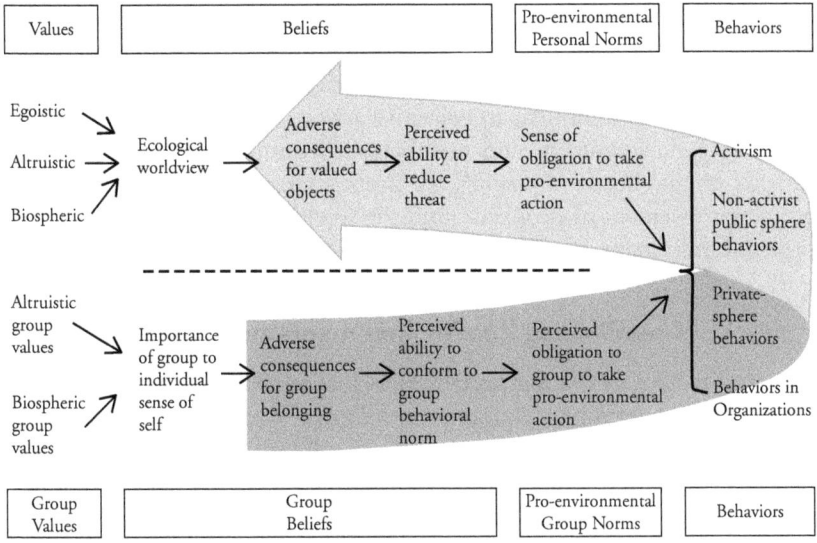

Figure 11.5   Collective identity – value-belief-norm theory

amongst these cohorts seemed to be as much about interactions within the group as any external target of protest. These belonging behaviors seemed to Fraser to suggest that group affinity and conformance with group norms may also predict pro-environmental behaviors.

Fraser's qualitative studies of volunteers revealed that one major driver of engagement in activism and public-sphere pro-environmental behaviors was a person's desire to act in ways likely to be deemed consistent with the public performance expectations of a trusted group. This finding suggested that understanding and enacting group-normative conservation behaviors can be simultaneous processes that occur when one begins acting like the other members of a group. Recognizing that he was seeing a pathway that flowed in the opposite direction as the original VBN model (see Chapter 10), Fraser developed the collective identity – value-belief-norm theory (CI–VBN; see Figure 11.5), which suggests that group affinity and affirmation are likely to predict environmental behaviors, and lead over time to shifts in beliefs and values that become more pro-environmental than they were before. Fraser's research has not disproven VBN, and both theories highlight the accrued importance of small steps, but CI–VBN offers another explanatory path to action that captures the consequences of participation in group norm behavior on a person's prior beliefs and values.

Zoo and aquarium staff and volunteers have emerged in recent years as a powerful, action-oriented cohort. These are folks for whom zoo experiences are intricately tied to place-based identities and shared values that ground a firm commitment to collective action. Is there reason to believe that zoo staff, volunteers, and members who similarly identify with zoo mission might effectively magnify the goals, strategies, and successes of the growing movement of people around the world who seek to preserve and protect biodiversity, in order to build movement momentum and power?

Coordinating Zoo Mission Actors and Action

Researchers have shown that socialization at zoos can contribute to positive regard for community[35] and provide valuable out-of-home experiences that help establish children as part of a community.[36] It is also the case that zoo visitors are more demographically representative of community demographics than visitors to other museum types,[37] though we have noted that zoo visitors are psychographically distinct in important, mission-related ways. So, what is known about the role of zoos in the inspiration and coordination of conservation action across communities?

We demonstrated in Chapter 5 that fun is often conducive to joint affect and group cohesion, but how can zoos facilitate coordinated conservation efforts that bring zoogoers together around – and keep them committed to acting in keeping with – a shared ethic and common foci? Zoo leaders can start by learning from how zoos are experienced and valued by the two cohorts most directly associated with zoo mission, beyond staff: zoo volunteers and zoo members.

For decades, conservation psychologists around the world have been applying research findings beyond the individual level to learn more about the spheres of community and cultural beliefs, values, and behaviors, seeking to establish how positive and negative experiences with nature influence social norms. Value attributed to biodiversity and the healthy biosphere on which living things depend, as well as self-aware understandings of the connectedness of the human and the natural, appear to be conditions that support purposeful collective action for conservation.

---

[35] Longhurst et al. 2004.   [36] Chase 1993; Carter 2003.
[37] The AZA maintains an annual update on zoo and aquarium demographics, data typical of what is found by most of the international zoo and aquarium associations: www.aza.org/partnerships-visitor-demographics.

Several studies have shown that some people invest a great deal of their self-concept in the zoo, valuing the zoo as an important place with an organizational mission that they feel inextricably tied to. A Wildlife Conservation Society study, for example, showed that sense of personal identification with the zoo and the zoo's conservations efforts grew increasingly stronger among individuals who remained members for many years, with values of visitation and the zoo as a venue decreasing over time. Furthermore, the same study revealed that zoo volunteers, who had chosen to work in service of the mission and identify themselves as part of the institution, demonstrated extremely robust identities *as* zoo volunteers.[38]

A 2009 study revealed substantial unrecognized potential of these groups to serve as social vectors for conservation information.[39] That study found that zoo volunteers were primarily tasked with leading visitor tours of the zoo and delivering education about the collections to school groups, or impromptu visitor interpretation throughout the park. These volunteers identified more with their zoo volunteer cohort than with their prior occupations, which were as wide ranging as cardiac surgeon, school bus driver, mechanic, and stay-at-home parent. Yet, their experiences building perceived relationships with the zoo animals and the shared affinity they had with other volunteers who shared their values for wildlife protection built among them a strong collective identity. Biweekly experiences together reinforced their sense of self as a conservationist. After a few years of volunteer activities and the social reinforcement they received from their confreres, zoo volunteers in the study became substantially more likely to engage in conservation-focused activism in their communities. They engaged their social networks at golf clubs, bridge clubs, boating activities, and other social contexts in conservation discussions. They advocated for environmentally responsible voting.

That study and a follow-up national survey[40] demonstrated that in-group experiences within the zoo volunteer cohort directly influenced the degree to which volunteers would engage in pro-environmental activities in their private lives, public lives, and organizations to which they belonged, and increased the likelihood they would become public activists for wildlife conservation. At this writing, accredited AZA institutions report that they have over 164,000 volunteers who offer 7.6 million hours of service each year.[41] These recent studies show, however, that this on-site

---

[38] Fraser and Sickler 2009.   [39] Fraser et al. 2009.   [40] Fraser 2009b.
[41] www.aza.org/from-the-desk-of-dan-ashe/posts/grateful-for-volunteers.

volunteer work is just the tip of the iceberg representing the environmental social activism zoo volunteers undertake in their lives overall.

Communities of Practice

After more than a decade of work, NNOCCI has become a robust social network of museum educators from settings such as zoos and aquariums that supports collective and individual conservation action in the United States.[42] At last count, group members embedded in leadership positions at more than 180 zoos, aquariums, and nature centers were actively coordinating their messaging. Moreover, more than 38,000 professionals have participated in NNOCCI training to employ effective climate interpretation strategies to activate public engagement in climate solutions. While supporting both self-directed personal behavioral change amongst trainees and their individual participation in group-based activities has become central to NNOCCI efforts, the collective has also been very intentional about fostering a vibrant community of practice.[43]

The NNOCCI training model emphasizes civic activities to address climate change that are likely to magnify the scope of impact achievable by solitary committed individuals. NNOCCI training seeks to equip professionals whose work involves interpreting climate science with communication and engagement strategies that feature empowering options. Such pathways might include, for example, strategies for initiating constructive conversations with friends, family members, and work colleagues, as well as for participating in and drawing attention to community activities that support energy production and consumption practices that are not based on fossil fuels.

Studies of environmental groups have shown that coherence can descend into infighting and dysfunctional behaviors if emotional stressors overwhelm the center of the community.[44] Emotional support is a key aspect of NNOCCI training, provided to help participants see themselves as part of a social cause and wider network in their workplace roles as interpreters who explain climate science to public audiences. This integrated approach to movement formation and sustained growth thus offers

---

[42] The study of NNOCCI social network design and cohesiveness is reported in Spitzer et al. 2020.

[43] The NNOCCI community continues to monitor and assess the growth and coherence of the network based on the work of Etienne Wenger and colleagues. See Wenger 1998 for an overview of the theory and Wenger et al. 2011 for the assessment rubrics used by that organization.

[44] Saunders 2008.

a supportive social network and lifelong learning model of direct relevance to the shared values and commitments of the group.

Together, network leaders and NNOCCI training program alumni are working to encourage people to accept that their concerns about climate science are normal, not exaggerated, and that agreement evident across and beyond the United States should encourage them to engage in public sphere behaviors to demonstrate their shared concerns. In organizations, groups, and forums that they value, members of the NNOCCI network are encouraged to initiate and support enriched discussions about creative solutions specifically relevant to the spheres of influence of the social networks represented therein. The goal is to identify and spread ideas and motivation regarding actions that may be salient to other tiny publics well positioned to collectively pursue and implement community- and regional-level changes in norms and policies. NNOCCI-trained climate change communicators speak regularly with groups visiting zoo and aquarium settings, where they know visitors overall to be active learners who are potential conservation allies or already-engaged environmental movement actors seeking pathways to deeper engagement.

One interesting pattern that has emerged through this work is that people who feel committed to climate change activism tend to diminish the degree to which they believe others are comparatively concerned about climate change. This perceived lack of concern suppresses their willingness to engage in discussions about climate change that might activate broader and deeper responses to climate change–related threats. This pattern points to a tacit social belief that concern for climate change should be or needs to be diminished in social settings; a belief that is a psychological inhibitor of action. Activism can stall when perceived or experienced negative social sanctions threaten individuals' sense of community, and because innovative climate change solutions cannot emerge and take hold without conversations about climate science and action, NNOCCI recognizes the importance of overcoming this barrier preventing those passionate about the protection and preservation of the biosphere from coming together and inviting others to act on shared values.

At present, United States citizens (and individuals in many other places) find themselves uncomfortably bound by often-polarized discourses, where many of the loudest voices seem either extremely concerned or in denial of obvious and overwhelming evidence of environmental decline. Such conditions make it challenging for those in between to initiate or engage in conversations that touch on climate change. Sociologist Kira Norgaard has suggested that social codes of conduct in the United States frame polite conversation as casual, light, and nonthreatening, which often means people steer toward local issues and any matter unlikely to avoid raised eyebrows.[45] To begin to identify and practice moving through these norms and perceived barriers to advancing conversations about environmental threats and solutions, the NNOCCI collective has confirmed the foundational importance of reasoned conversation around setting achievable goals in small groups as a central goal for all climate change and ocean science interpretation work.

Studies of collective identity, the degree to which an individual's perceived self-worth is inextricably tied to a group they value or feel emotionally attached to, suggest that social group membership influences the likelihood that an individual will tend to seek to behave as a typical group member. As such, active group participation and co-development of ideas and plans makes it more likely that group members will take similar actions in the immediate present and in the longer term. Use of a high-touch model that incorporates ongoing connections, lifelong learning, and social–emotional support are specific conditions that appear to increase the likelihood that committed conservationists become advocates in spheres that lie within *and beyond* the site where their identity is reinforced and magnified: the zoo.[46]

While zoo and aquarium leaders have tended to focus on individual experiences within the zoo and debated the extent to which a single visit can lead to individual-level action, the NNOCCI movement is an example of a high-impact expression of thoughtfully coordinated collective action.

Where Fairfield Osborn and other zoo-associated conservation activist leaders of the past adapted available forms of communication to reimagine field guides, literature, and public commentary as a sociopolitical engine for conservation that transcended empire, academic discipline, and interest group, the affordances of interconnectivity have now opened new possibilities for outreach by zoo-associated activists with numerous roles in these institutions to engage public audiences with new information, ideas, and opportunities for engagement.

The Social Radiation Approach

To maintain genetic diversity in their collections, provide adequate space for animals across their life course, and address the needs of geriatric animals, zoo management is a cooperative enterprise. The result of the

---

[45] Norgaard 2011.   [46] Swim et al. 2014.

highly technical demands of zoo operations and species management has driven these institutions to establish and maintain a very tight community of practice that has few parallels in other types of museum or social service provision. Accredited zoos are, in effect, a massive global cooperative that has developed skills and tools for sharing all aspects of their practice. Education programs being delivered in zoos around the world represent remarkable consistency in design and execution. In the United States, United Kingdom, European Union, Canada, and Australia, there is shared agreement on mission, goals, and objectives for local zoo education, highlighted by shared practices published in select journals.[47]

The NNOCCI program was created to support that larger community of practice with a tight-knit subgroup that can work toward a common purpose. At the outset, research leads John Fraser and Janet Swim developed a national survey to assess conservation and civic engagement in conservation issues, stratifying their study to determine whether there was a difference between the patterns associated with zoos that were already engaged in climate change communication and zoos that were not. Study data suggested that zoogoers were distinct from the general population, and that zoos that are more engaged in environmental messaging also attract visitors who are more likely to be engaged, overall (see Figure 11.6). This suggests that zoos have good reason to choose to be more activist in their messaging – and will be likely to draw visitors who have a greater likelihood of being vectors for social change, if they do so.

Based on the degree of trust and authority local communities bestow upon zoos and aquariums and NNOCCI's unique position as a very robust community of practice, a second wave of research will be undertaken in the 2020s regarding the role of zoos as community catalysts that can convene and coordinate more direct local action. Experiments currently underway at this writing are exploring the degree to which zoos can provide service as community organizers for local environmental action groups. This work requires new skills that do not map easily to the traditional role of "zoo educator" but show promise to produce new and effective opportunities to pursue and achieve zoo mission.

An activist community with global reach, NNOCCI is promoting widespread action through public voice and looking at bilateral support to agitate for short-term strategies that can produce, replicate, and scale

---

[47] Major journals supporting the zoo education community include *Curator: The Museum Journal, Visitor Studies, International Zoo Yearbook, Journal of Zoo and Aquarium Research, Zoo Biology,* and *International Zoo Educators' Journal.*

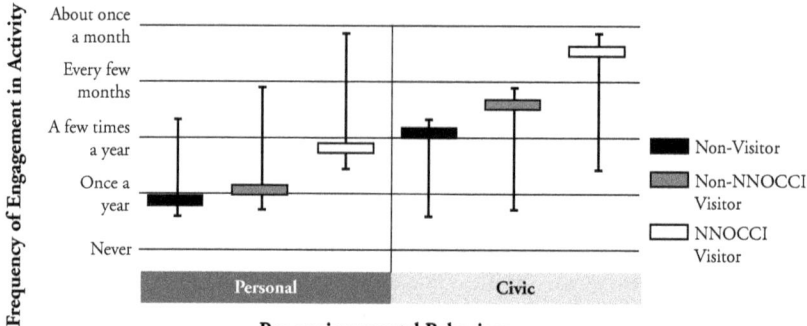

Figure 11.6   Variation in public engagement in personal and civic environmental behaviors and frequency of reported visitation to zoos and aquariums

conservation impacts that matter. It is possible to leverage that power more aggressively. Hundreds of zoos are represented by the participant action learning communities that comprise the NNOCCI network. Offering hope and long-term capacity value, these trained climate change interpreters are returning to the zoos that employ them equipped with tools, affirmed shared identities, and emotional support to fortify them in their important work and increase their effectiveness. This model should yield ongoing impact because it is rooted in the creation of a zoo-affiliated social community of highly committed conservation activists willing and able to use context-relevant aspects of best practice to deepen the conservation ethic and commitment to action of potential allies in their immediate and online spheres.

Data associated with many of the studies described throughout this book point to the value of a new vision of emotional *and* science-grounded strategies for connecting with public audiences to better advance wildlife care and conservation in as many communities and societies as possible. This new community-based approach envisions tiny public goals and engagement strategies that have little overlap with traditional approaches to activating zoo visitors through natural history education and appeals of denial. We suggest that zoos have been under-leveraged, but are ideally situated to create conditions under which groups and communities can participate in joint meaning-making that is consistent with a "non-persuasive" approach.[48] They can use techniques such as promoting emotional engagement to build relational understandings of causes and consequences

[48] Fischhoff 2007.

rather than experts advocating particular policies or actions.[49] In lieu of defaulting to the education of youth for change in/by the next generation, we echo the suggestion of other scholars that zoos have the authority and coordination skills to convene, motivate, and coordinate civic groups to collaborate on community-level solutions that are socially acceptable, feasible, and effective.[50]

When shared and actionable perspectives gain momentum and take the form of a social movement, efforts and outcomes can become amplified and may include shifts in popular discourse; the production and pursuit of ideas and goals; and success pressuring actors at the organizational, local, regional, industry, national, and/or global levels.[51] Zoo users have mission-relevant psychographic distinctions from their neighbors, and zoos have the authority and potential to focus on mission-related concerns and solutions of value to the community. As such, the unique assets and affordances of zoos and aquariums can and should support the multimodal engagement of concerned community members in deliberation, dialogue, and collaboration around priority conservation concerns.

---

[49] A full argument for this future research is outlined in Spitzer et al. 2020.
[50] Levine 2007; Rudolph and Horibe 2015.   [51] Dryzek 1997.

CHAPTER 12

Integration
The Socially Valuable Zoo

In the mid-1980s, the leader of zoo education at a zoo that had opened as a shining international exemplar of modern zoo design just twenty years prior convened a meeting to set out the goals for the first major exhibit improvements since the zoo's opening. As the meeting started, he made a point-blank statement that he would be satisfied if people felt a sense of awe at the wonder of nature when they encountered the redesigned exhibits. Aghast, John Fraser, the newly minted architect charged with leading the renovation, blurted out, "That's it?! That's what we're doing with hundreds of thousands of dollars in building costs?" While that memorable meeting went very poorly, it did inspire and motivate John toward a career-long pursuit of better questions and answers regarding why zoos matter to their users.

At the turn of the twentieth century, zoos in very diverse locales came to agreement about a common purpose that leading institutions such as the London Zoological Society, the Smithsonian's National Zoo, and the Wildlife Conservation Society (formerly the New York Zoological Society) had envisioned at their founding: Zoos aimed to help people better understand the natural world. Over the next hundred years, research showed that zoos needed to lead the charge to save biodiversity and the habitats wildlife depends on, and zoos had demonstrated that they could do much more than monitor and build awareness about species decline. A collective conservation mission was adopted by accredited zoos everywhere in the world.

Leading zoos have taken direct responsibility for species recovery and conservation action targeting species as different as bison, butterflies, and great apes. Field conservation work has given zoos renewed authority to advocate for more conservation action by the communities in which they are based. The last twenty years of social science research has disproven the fallacy that zoos are nothing more than shallow entertainment venues and confirmed that zoo users are deeply engaged with conservation goals.

It is important to acknowledge that, with respect to analyzing the pursuit of zoos' institutional mission of conservation, the social sciences have received much less focus and funding than the biological sciences. While tentative research into mission-related social phenomena in the late twentieth century has remained underutilized, today we see great opportunity and fresh reasons for zoos to build on the concepts and foundational studies we covered in this book.

The environmental challenges facing society are great. Zoos and aquariums have unique social affordances that situate these institutions and those who work in them as distinctly well positioned for advancing an active conservation ethic in communities all over the world. Zoos are places for pleasure and personal meaning-making, places where people go to act on their personal values and ethics or to pass those values on to the next generation. They are places to fulfill the desire to feel part of the nature continuum and explore social belonging and identity. They are also places to ruminate on social attitudes and collective action pathways. All these happenings are not only consistent with, but of crucial importance to, the advancement of a conservation agenda.

Foundational understandings about how the zoo is situated in and through mental process now allows conservation organizations to test and put into action new communications strategies that more effectively motivate action. Our goal was to highlight many such findings, in the hope that many of the insights we share in this book might be applied to the development or adaptation of exhibits and programs that can be effectively evaluated for the extent to which new approaches appear to advance mission-aligned understandings, attitudes, and actions. Because learning is situated *within the learner*, rather than something done *to the learner*, our focus has been on the affordances we see when we study the zoogoing phenomenon, and the implications of these findings for the pursuit and achievement of mission goals.

To delegitimize the anti-zoo movement *and* reposition for greater mission impact, zoos and aquariums must reconsider the identity they want the public to perceive, their learning priorities, and how staff and volunteers from every department across the institution can strengthen mission-supporting efforts. We see these foci as integral for institutions to address, in their efforts to connect with audiences and pursue mission goals. While people use zoos with their own priorities and motivations, zoos are social settings and represent a vast and trusted social enterprise, which makes these sites a key vector for changing how animals and the environment are viewed, valued, and engaged with, by all of us, through action.

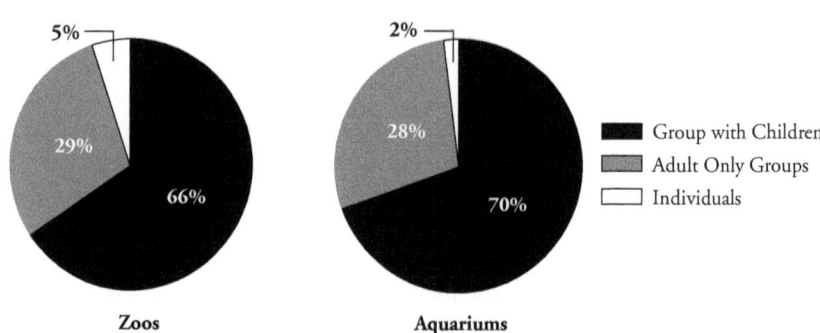

Figure 12.1   Average distribution of adult-only groups, adult-with-children groups, and individuals entering US zoos and aquariums

Zoos need to focus more strategically on adult audiences. A solid one-third of all visits do not include children (see Figure 12.1), and even when children *are* present, adults are actively engaging them in discussions about what is valued and what is right to do (a pattern we discussed in Chapters 3, 4, and 7).

Zoo users want to connect. They want to connect with each other, with animals, with opportunities to advance conservation, and with zoo staff and volunteers trusted and able to help them interpret the animals, experiences, and mission-related concepts they encounter at the zoo. As new strategies to connect zoogoers with affirming, identity-building, and community-building conservation messaging and opportunities for con-servation action continue to be fine-tuned in leading zoos around the world, emergent best practices can be drawn from.

It is fascinating, and relevant, to consider the role zoos play in the mind, where they exist as a mnemonic for the uncontrolled chaos of our world and also as a community symbol of the degree to which we care for nature. The zoo is a social actor whose communications within the parks and through public media shape values about conservation action. When visitors arrive, they have already assigned the zoo responsibilities of ensur-ing animal well-being, doing conservation work, and being a good collab-orator and leader in the conservation domain. Each visit tests whether the zoo is living up to expectations and is used to think about what more can be done. Importantly, too, zoo visits often become memorable moments in a person's lifelong process of integrating new and old experiences and understandings into a set of values.

When we acknowledge that zoo visiting is a continuum of experiences, each visit a node in a learning journey, we put less importance on each instance achieving behavior change. We can instead think of each visit, online connection, and media message as another touchstone. Using findings reported here, zoos can deepen their social value and impact by creating more continuity across these points of reference. Zoo *users* are the vectors that carry conservation actions and momentum into the larger community. Visitors must thus be situated as partners working with zoo staff and a broader worldwide movement to achieve shared goals.

Thoughtful and creative strategists can use these data to craft more engaging learning opportunities for influencers and allies from all spheres. Zoos have the capacity to organize community resources and activist groups toward common conservation goals. In addition to interweaving on-site conservation messaging with thought- and conversation-provoking design strategies and field conservation endeavors that capture media and public attention, coordinated action in the conservation sphere can include pressuring state legislatures and using the bully pulpit of public trust to encourage local industry partners to pursue solutions to existing, emerging, and inevitable environmental threats.

Given the authority zoos wield as sites visited by notably large and diverse audiences, and the perceived moral authority of zoos and zoo staff with respect to animal care, larger zoos in many countries have long been active in the political sphere. Indeed, many politicians worry about the power and voice of their local zoo, situating the zoo as a civic actor capable of drawing attention to community leaders who are not doing what they ought.[1]

Our completion of this book coincided with the closure of the world's zoos in response to COVID-19, a pandemic that has threatened and taken the lives of people in all corners of the world. While this respiratory disease is unlike anything seen in over a century, the nineteenth coronavirus type, like all zoonotic diseases of the past, jumped into the human population as a direct result of the irresponsible management of wild or domesticated animals in the global food web. Recent examples have included SARS, the 2009 H1N1 influenza pandemic, avian influenzas H5N1 and H7N9, and the Middle East respiratory syndrome coronavirus known as MERS-CoV. These outbreaks have highlighted crucial intersections of human health, human interactions with animals, and conservation advocacy that require thought leaders to bridge the too-often siloed domains of biological and zoological science, psychology, communication, education, policy, and practice.

---

[1] Fraser and Sickler 2008b.

Zoos and zoo research are clearly much needed. We have shown throughout this book that ongoing research, applied expertise, and activism in these areas is very directly connected to zoo operations and zoo mission. Zoos have also long been at the forefront in drawing attention to risks and costs of wildlife trafficking and recommending that the public act and apply pressure to end these life-threatening practices. This pandemic will not be the last, and zoos have important work to do. Zoos help public audiences experience the wonder of nature, use emotion to understand public concerns, and identify steps to be part of establishing norms that sustain what remains of global biodiversity and protect human populations.

Through their work as an international collective, zoos no longer need to see themselves as working autonomously. Zoos must embrace their shared global voice and work creatively together to provoke common understandings and goals, and collective action, in local, regional, and global contexts. By drawing their diverse users into the wildlife conservation movement and allocating resources to raise and amplify the voices and actions of movement supporters, behavior and policy change can be accelerated across the globe.

While we have offered a wide-ranging view of how zoos are situated in the public mind, and the very unique (and potentially high-impact) affordances of these settings, we caution that the dominant trend in zoos' exhibitions and public communications continues to be narrow focus on why environmental problems have arisen and why environmental stewardship matters. This trend is problematic because it arises from an assumed lack of knowledge, rather than situating zoogoers as individuals and groups who already share the zoo's conservation values and are potential allies and colleagues willing and motivated to seek change if a range of clear pathways are presented. We hope this book will be a helpful starting point for deeper dialogue about how zoos can more fully realize the institutional power of their trusted public voice, programs, spaces, staff, volunteers, members, and live animal exhibits to engage people who share values around and emotional connections to animals as changemakers committed to immediate and ongoing individual and collective conservation action.

Critical to this potential is the redirection and clever use of staff and volunteers; these are trusted professionals with unique opportunities to leverage the management and care of live animals and the context at hand to scaffold learning that integrates sensory, cognitive, emotional, and social stimuli as multimodal resources that help zoogoers connect and make meaning. Human resource capacities are inevitably and deeply interwoven with zoo mission success because the social, institutional, and historical contexts of these settings impact, and are reshaped by, the scripted and unscripted human, animal, and interspecies interactions that occur therein. It is thus unsurprising that knowledgeable staff who are authorized and equipped to help users interpret their on-site experiences tend to deepen engagement and improve learning outcomes in tangible and meaningful ways.

Moral and relational practices, essential cruxes of conservation, always remain fluid, unfinished, and complex. Sociologist Donald Gibson has pointed out that real concern for the environment includes full and open discussion of the assumptions and implications of numerous ideas, a focus on specific problems, and solution-seeking that prioritizes human development and well-being. Though Gibson makes the legitimate point that economic, political, and social self-interest has often been disguised as concern for and protection of the environment, we reject his characterization of "environmentalism" as a reactionary worldview of the global upper class, and strongly disagree that sincere solution-seeking must be "dispassionate and objective."[2]

Indeed, we know that hedonic and eudaimonic well-being are valued and valuable outcomes that can accrue when people have access to personal encounters with live animals, and firmly believe that such experiences are not inconsistent with, and can even propel, a conservation agenda. The next generation of the global zoo movement need not apologize for learning from captive animals. It is both acceptable and beneficial to connect and find pleasure in the company of wild beings, to embrace those beings as full selves that present opportunities to learn through and about fear, affinity, scent, visual variety, relational dynamics, and instinctive behavior. We respond to such experiences on multiple levels, even when the context is managed and designed by humans.

Threats to biodiversity, natural spaces, and natural systems have complex causes and implications, but zoos must come to terms with the fact that social science research is central to and necessary for making conservation a reality.

Informed by social science research, the conservation agenda of the future zoo will:

- demonstrate inclusive and democratic characteristics, engaging users from all walks of life as movement allies;
- value dialogue, passion, science, and new technologies as critical components of new solutions for identifying, understanding, and mitigating conservation threats;

---

[2] Gibson 2014.

- affirm and honor the importance of human well-being; and
- situate community and cultural norms as sources of, as well as foundations for potential solutions to, conservation concerns and obstacles.

The contemporary zoo offers much in terms of social value and has the capacity to achieve conservation mission impact. The challenge now is to stop undervaluing the latent potential of the form. Narrow emphasis on individual-level, single-visit, scripted knowledge transfer impacts and metrics have obscured how public audiences engage, and hope to engage, with these institutions. Such lenses, coupled with inward-facing preoccupations regarding the strengths and weaknesses of one's own zoo, also continue to limit the ability of zoos to move toward becoming a truly global movement optimizing their collective reach and impact. While the form has the capacity and mission call to make this happen, it cannot occur unless zoos everywhere move quickly to acknowledge and better leverage the many ways these institutions are perceived, valued, and used.

References

Abrams, Dominic, and Michael A. Hogg. 2004. Metatheory: Lessons from social identity research. *Personality and Social Psychology Review* 8, no. 2: 98–106.

Acampora, Ralph. 2001. Real animals? An inquiry on behalf of relational zoöntology. *Human Ecology Review* 8, no. 2: 73–78.

———. 2005. Zoos and eyes: Contesting captivity and seeking successor practices. *Society and Animals* 13, no. 1: 69–88.

———. 2006. *Corporal compassion: Animal ethics and philosophy of body.* Pittsburgh, PA: University of Pittsburgh Press.

Ajzen, Icek. 1991. The theory of planned behavior. *Organizational Behavior and Human Decision Processes* 50, no. 2: 179–211.

Allen, Sue. 2002. Looking for learning in visitor talk: A methodological exploration. In Gaea Leinhardt, Kevin Crowley, and Karen Knutson, eds., *Learning conversations in museums*, 259–303. Mahwah, NJ: Lawrence Erlbaum Associates, Inc.

Andereck, Kathleen L., and Linda L. Caldwell. 1994. Variable selection in tourism market segmentation models. *Journal of Travel Research* 33, no. 2: 40–46.

Anderson, Ursula S., Angela S. Kelling, Robin Pressley-Keough, Mollie A. Bloomsmith, and Terry L. Maple. 2003. Enhancing the zoo visitor's experience by public animal training and oral interpretation at an otter exhibit. *Environment and Behavior* 35, no. 6: 826–841.

Anthony, Lawrence, and Graham Spence. 2009. *The elephant whisperer: My life with the herd in the African wild.* New York: Henry Holt and Company.

Archibald, Robert R. 2004. *The new town square: Museums and communities in transition.* Walnut Creek, CA: AltaMira Press.

Ash, Doris. 2003. Dialogic inquiry in life science conversations of family groups in a museum. *Journal of Research in Science Teaching* 40, no. 2: 138–162.

Ashmore, Richard D., Kay Deaux, and Tracy McLaughlin-Volpe. 2004. An organizing framework for collective identity: Articulation and significance of multidimensionality. *Psychological Bulletin* 130, no. 1: 80–114.

Associated Press. 2005. Joy as New Orleans zoo opens again, recovered from Katrina. *Pittsburgh Post-Gazette*, November 26.

Australian Bureau of Statistics. 2004. *Measuring social capital, an Australian framework and indicators.* Canberra, Commonwealth of Australia.

Axelsson, Tony, and Sarah May. 2008. Constructed landscapes in zoos and heritage. *International Journal of Heritage Studies* 14, no. 1: 43–59.

Ballantyne, Roy, and Jan Packer. 2011. Using tourism free-choice learning experiences to promote environmentally sustainable behaviour: The role of post-visit "action resources." *Environmental Education Research* 17, no. 2: 201–215.

2016. Visitors' perceptions of the conservation education role of zoos and aquariums: Implication for the provision of learning experiences. *Visitor Studies* 19: 193–221.

Ballantyne, Roy, Jan Packer, and Lucy A. Sutherland. 2011. Visitors' memories of wildlife tourism: Implications for the design of powerful interpretive experiences. *Tourism management* 32, no. 4: 770–779.

Ballantyne, Roy, Karen Hughes, Julie Lee, Jan Packer, and Joanne Sneddon. 2018. Visitors' values and environmental learning outcomes at wildlife attractions: Implications for interpretive practice. *Tourism Management* 64: 190–201.

Balmford, Andrew, Nigel Leader-Williams, Georgina Mace, Andrea Manica, Olivia Walter, Chris West, and Alexandra Zimmermann. 2007. Message received? Quantifying the impact of informal conservation education on adults visiting UK zoos. In Alexandra Zimmermann, Matthew Hatchwell, Lesley Dickie, and Chris West, eds., *Zoos in the 21st century: Catalysts for conservation?* Cambridge: Cambridge University Press.

Bandura, Albert. 2012. Moral disengagement. In Daniel J. Christie, ed., *Encyclopedia of peace psychology*. Hoboken, NJ: John Wiley and Sons, Inc.

Baratay, Eric and Elisabeth Hardouin-Fugier. 2004. *Zoo: A history of zoological gardens in the West*, second ed. Trans. Oliver Welsh. London: Reaktion Books.

Baumeister, Roy F., ed. 1999. *The self in social psychology*. London: Psychology Press.

Baumeister, Roy F., and Mark R. Leary. 1995. The need to belong: Desire for interpersonal attachments as a fundamental human motivation. *Psychological Bulletin* 117, no. 3: 497–529.

Beardsworth, Alan, and Alan Bryman. 2001. The wild animal in late modernity: The case of the Disneyization of zoos. *Tourist Studies* 10, no. 1: 83–104.

Beckman, John. 2014. *American fun: Four centuries of joyous revolt*. New York: Pantheon.

Beirne, Pierce, and Nigel South. 2012. Animal rights, animal abuse and green criminology. In *Issues in green criminology*, 77–106. Philadelphia: Taylor and Francis.

Bekoff, Marc. 2003. Minding animals, minding earth: Science, nature, kinship, and heart. *Human Ecology Review* 10, no. 1: 56–76.

Bekoff, Marc, and Jessica Pierce. 2010. *Wild justice: The moral lives of animals*. Chicago: University of Chicago Press.

Benbow, S. Mary P. 2004. Death and dying at the zoo. *Journal of Popular Culture* 37, no. 3: 379–384.

Benbow, S. Mary P., and Bonnie C. Hallman. 2008. Reading the zoo map: Cultural heritage insights from popular cartography. *International Journal of Heritage Studies* 14, no. 1: 30–42.

Berger, John. 1980. *Why look at animals? About looking.* New York: Pantheon.

Berkovits, Annette Libeskind. 2017. Confessions of an accidental zoo curator. *Curator: The Museum Journal* 60, no. 3: 267–271.

Birney, Barbara A. 1988. Criteria for successful museum and zoo visits: Children offer guidance. *Curator* 31, no. 4: 292–316.

Bishop, Patrick. 2004. Haven for zoo animals provides an escape from mayhem of city. *Telegraph*, March 20.

Bloodhart, Brittany, Janet K. Swim, and Matthew J. Zawadzki. 2013. Spreading the eco-message: Using proactive coping to aid eco-rep behavior change programming. *Sustainability: Special Issue on Psychological and Behavioral Aspects of Sustainability* 5: 1661–1679.

Bodamer, M. D., and J. M. Sankovic. 2000. "We're all cousins!" A sampling of public comments at a zoo, reflecting people's sibling relationship with chimpanzees. *Zoo Biology* 18, no. 5: 443–448.

Boltanski, Luc, and Eve Chiapello. 2006. *The new spirit of capitalism.* New York: Verso.

Bott, Suzanne, James G. Cantrill, and Olin Eugene Myers, Jr. 2003. Place and the promise of conservation psychology. *Human Ecology Review* 10, no. 2: 100–112.

Bovens, Mark, Thomas Schillemans, and Robert E. Goodin. 2014. Public accountability. In Mark Bovens, Thomas Schillemans, and Robert E. Goodin, eds., *The Oxford handbook of public accountability*, 1–20. New York: Oxford University Press.

Braverman, Irus. 2013. *Zooland: The institution of captivity.* Stanford, CA: Stanford University Press.

2015. *Wild life: The institution of nature.* Stanford, CA: Stanford University Press.

Brewer, Marilynn B. 2001. The many faces of social identity: Implications for political psychology. *Political Psychology* 22, no. 1: 115–125.

Bruni, Coral M., John Fraser, and P. Wesley Schultz. 2008. The value of zoo experiences for connecting people with nature. *Visitor Studies* 11, no. 2: 139–150.

Burghardt, Gordon M. 1996. Environmental enrichment or controlled deprivation. In G. M. Burghardt, J. T. Bielitzki, J. R. Boyce, and D. O. Schaeffer, eds., *The well-being of animals in zoo and aquarium sponsored research*, 91–101. Greenbelt, MD: Scientists Center for Animal Welfare.

2013. Environmental enrichment and cognitive complexity in reptiles and amphibians: Concepts, review, and implications for captive populations. *Applied Animal Behavior Science* 147: 286–298.

Calvillo, Dustin P., and Whitney C. Hawkins. 2016. Animate objects are detected more frequently than inanimate objects in inattentional blindness tasks independently of threat. *Journal of General Psychology* 143, no. 2: 101–115.

Carbonell, Bettina Messias, ed. 2004. *Museum studies.* Oxford: Wiley-Blackwell.

Carr, David. 2011. *Open conversations: Public learning in libraries and museums.* Santa Barbara, CA: ABC-CLIO, LLC.

Carter, Susanne. 2003. *Educating our children together: A sourcebook for effective family-school-community partnerships.* Eugene, OR: Consortium for Appropriate Dispute Resolution in Special Education.

Charmaz, Kathy. 2000. Grounded theory: Objectivist and constructivist methods. In Norma K. Denzin and Yvonna S. Lincoln, eds., *Handbook of qualitative research.* London: Sage Publications.

Chase, Richard Allen. 1993. Alternative learning networks. *Children's Environments* 10, no. 2: 136–163.

Clark, Fay E. 2013. Marine mammal cognition and captive care: A proposal for cognitive enrichment in zoos and aquariums. *Journal of Zoo and Aquarium Research* 1: 1–6.

Clayton, Susan. 2003. Environmental identity: A conceptual and an operational definition. In Susan Clayton and Susan Opotow, eds., *Identity and the natural environment: The psychological significance of nature.* Cambridge, MA: MIT Press.

Clayton, Susan, and Amara Brook. 2005. Can psychology help save the world? A model for conservation psychology. *Analysis of Social Issues and Public Policy* 5: 87–102.

Clayton, Susan, and Gene Myers. 2009. *Conservation psychology: Understanding and promoting human care for nature.* West Sussex: Wiley-Blackwell.

Clayton, Susan, Jerry Luebke, Carol Saunders, Jennifer Matiasek, and Alejandro Grajal. 2014. Connecting to nature at the zoo: Implications for responding to climate change. *Environmental Education Research* 20, no. 4: 460–475.

Clayton, Susan, John Fraser, and Claire Burgess. 2011. The role of zoos in fostering environmental identity. *Ecopsychology* 3, no. 2: 87–96.

Clayton, Susan, John Fraser, and Carol Saunders. 2009. Zoo experiences: Conversations, connections, and concern for animals. *Zoo Biology* 28, no. 5: 377–397.

Clayton, Susan, and Khoa D. Le Nguyen. 2018. People in the zoo: A social context for conservation. In Ben A. Minteer, Jane Maienschein, and James P. Collins, eds., *The ark and beyond: The evolution of zoo and aquarium conservation.* Chicago: University of Chicago Press.

Clayton, Susan, and Susan Opotow, eds. 2003. *Identity and the natural environment: The psychological significance of nature.* Cambridge, MA: MIT Press.

Coe, Jon Charles. 1985. Design and perception: Making the zoo experience real. *Zoo Biology* 4, no. 2: 197–208.

1997. Entertaining zoo visitors and zoo animals: An integrated approach. Paper prepared for the 1997 AZA Annual Conference.

Conway, William. 1968. How to exhibit a bullfrog: A bed-time story for zoo men. *Curator: The Museum Journal* 11, no. 4: 310–318.

1994. Preface. In Linda Koebner, ed., *Zoo book: The evolution of wildlife conservation centers,* 13–15. New York: Tom Doherty Associates Inc.

Csikszentmihalyi, Mihaly. 1990. *Flow: The psychology of optimal experience.* New York: Harper & Row.

1997. *Finding flow: The psychology of engagement with everyday life.* New York: HarperCollins Publishers.

Dahlburg, John-Thor. 1996. Afghan capital's beleaguered zoo is a microcosm of nation. *Los Angeles Times,* October 11, p. A2.

Daly, Beth, and Suzanne Suggs. 2010. Teachers' experiences with humane education and animals in the elementary classroom: Implications for empathy development. *Journal of Moral Education* 39, no. 1: 101–112.

Dana, John Cotton, 1999. *The new museum: Selected writings.* Newark, NJ: Newark Museum Association.

Darier, Éric, ed. 1999. *Discourses of the environment.* Oxford: Blackwell Publishers Ltd.

Davey, Gareth. 2005. Is zoo-going a human instinct? Biophilia and zoos. *International Zoo News* 52, no. 8: 452–459.

2006. Visitor behavior in zoos: A review. *Anthrozoös* 19, no. 2: 143–157.

Denning, Kathryn. 2008. Regarding the zoo: On the deployment of a metaphor. *International Journal of Heritage Studies* 14, no. 1: 60–73.

DeVault, Marjorie L. 2000. Producing family time: Practices of leisure activity beyond the home. *Qualitative Sociology* 23, no. 4: 485–503.

Dewey, John, 1938/1963. *Experience and education.* London: Collier-MacMillan.

Dhont, Kristof, Gordon Hodson, Ana C. Leite, and Alina Salmen. 2019. The psychology of speciesism. In Kristof Dhont and Gordon Hodson, eds., *Why we love and exploit animals: Bridging insights from academia and advocacy.* Abingdon: Routledge.

Diamond, Jared. 2014. *The third chimpanzee for young people: On the evolution and future of the human animal.* New York: Seven Stories Press.

Dierking, Lynn D. 2005. Lessons without limit: How free-choice learning is transforming science and technology education. *Historia, Ciencias, Saude – Manguinhos* 12 (supplement): 145–160.

Dierking, Lynn D., Kim Burtnyk, Kirsten S. Buchner, and John H. Falk. 2002. *Visitor learning in zoos and aquariums: A literature review.* Silver Spring, MD: Association of Zoos and Aquariums.

Dietz, Thomas, and Paul C. Stern. 1995. Toward an individual model of choice: Socially embedded preference construction. *Journal of Socio-Economics* 24, no. 2: 261–279.

Dixon, John, and Kevin Durrheim. 2000. Displacing place-identity: A discursive approach to locating self and other. *British Journal of Social Psychology/British Psychological Society* 39 (part 1): 27–44.

Dohn, Niels Bonderup. 2013. Upper secondary students' situational interest: A case study of the role of a zoo visit in a biology class. *International Journal of Science Education* 35, no. 16: 2732–2751.

Dryzek, John S. 1997. *The politics of the earth: Environmental discourses.* New York: Oxford University Press.

Eagly, Alice, and Shelly Chaiken. 1993. *The psychology of attitudes.* Fort Worth, TX: Harcourt Brace Jovanovich College Publishers.

Egan, Kieran. 1997. *The educated mind: How cognitive tools shape our understanding*. Chicago: University of Chicago Press.

Ehrenfeld, David. 1995. Foreword. In Bryan G. Norton, Michael Hutchins, Elizabeth F. Stevens, and Terry L. Maple, eds., *Ethics on the ark: Zoos, animal welfare, and wildlife conservation*, xvii–xix. Washington, DC: Smithsonian Institution Press.

Falk, John H. 2005. Free-choice environmental learning: Framing the discussion. *Environmental Education Research* 11: 265–280.

Falk, John H., Joe E. Heimlich, and Kerry Bronnenkant. 2008. Using identity-related visit motivations as a tool for understanding adult zoo and aquarium visitors' meaning-making. *Curator: The Museum Journal* 51, no. 1: 55–79.

Falk, John H., Joe E. Heimlich, and Susan Foutz, eds. 2009. *Free-choice learning and the environment*. Lantham, MD: AltaMira Press.

Falk, John H., and Lynn D. Dierking. 1992. *The museum experience*. Washington, DC: Whalesback Books.

———. 2000. *Learning from museums: Visitor experiences and the making of meaning*. Walnut Creek, CA: AltaMira Press.

Falk, John H., Eric M. Reinhard, Cynthia L. Vernon, Kerry Bronnenkant, Joe E. Heimlich, and Nora L. Deans. 2007. *Why zoos & aquariums matter: Assessing the impact of a visit to a zoo or aquarium*. Silver Spring, MD: Association of Zoos and Aquariums.

Fine, Gary Alan, and Brooke Harrington. 2004. Tiny publics: Small groups and civil society. *Sociological Theory* 22, no. 3: 341–356.

Fine, Gary Alan, and Ugo Corte. 2017. Group pleasures: Collaborative commitments, shared narrative, and the sociology of fun. *Sociological Theory* 35, no. 1: 64–86.

Fischhoff, Baruch. 2007. Non-persuasive communication about matters of the greatest urgency: Climate change. *Environmental Science and Technology* 41: 7204–7208.

Fraser, John. 2004. Museums and civility. *Curator: The Museum Journal* 47, no. 3: 252–255.

———. 2006. Group identity, protest and evolution exhibits in America. *Museums and Social Issues* 1, no. 1: 85–100.

———. 2009a. The anticipated utility of zoos for developing moral concern in children. *Curator: The Museum Journal* 52, no. 4: 349–361.

———. 2009b. An examination of environmental collective identity development across three life-stages: The contribution of social public experiences at zoos. Ph.D. diss., Antioch University New England.

———. 2017. Thinking about museum type. *Curator: The Museum Journal* 60, no. 3: 263–265.

Fraser, John, and Carol Brandt. 2013. The emotional life of the environmental educator. In Marianne Krasney and Justin Dillon, eds., *Trading zones: Creating trans-disciplinary dialogue in environmental education*, 133–158. Oxford: Peter Lang Ltd.

Fraser, John, and Dan Wharton. 2007. The future of zoos: A new model for cultural institutions. *Curator: The Museum Journal* 50, no. 1: 41–54.

Fraser, John, David Wilkie, Robert Wallace, Peter Coppolillo, Roan Balas McNab, Lilian E. Painter, Peter Zahler, and Isabel Buechsel. 2009. The emergence of conservation NGOs as catalysts for local democracy. In Michael J. Manfredo et al., eds., *Wildlife and society: The science of human dimensions*, 44–56. Washington, DC: Island Press.

Fraser, John, Heather Lerner, John Voiklis, and Carol D. Saunders. 2020 (forthcoming). Alignment between zoo and aquarium missions and visitor values. In John Fraser, Joe E. Heimlich, and Kelly Riedinger, eds., *Understanding zoos and aquariums in the public mind*. New York: Springer Nature.

Fraser, John, and Jessica Sickler. 2008a. Conservation psychology: Who cares about the biodiversity crisis? In Eva Fearn, ed., *State of the wild 2008–2009: A global portrait of wildlife, wildlands, and oceans*, 206–212. Washington, DC: Island Press.

2008b. *Why zoos and aquariums matter: Handbook of research key findings and results from national audience survey*. Bronx, NY: Wildlife Conservation Society.

2009. Measuring the cultural impact of zoos and aquariums. *International Zoo Yearbook* 43, no. 1: 103–112.

Fraser, John, Joe E. Heimlich, and Victor S. Yocco. 2010. *Report number 20100226: American beliefs associated with increasing children's opportunities for experiences in nature*. Edgewater, MD: Institute for Learning Innovation.

Fraser, John, Sarah Gruber, and Kathleen Condon. 2008. Exposing the tourist value proposition of zoos and aquaria. *Tourism Review International* 11.

Fraser, John, Susan Clayton, Jessica Sickler, and Anthony Taylor. 2009. Belonging at the zoo: Retired volunteers, conservation activism, and collective identity. *Ageing and Society* 29, no. 3: 351–368.

Fraser, John, Victor Pantesco, Karen Plemons, Rupanwita Gupta, and Shelley J. Rank. 2013. Sustaining the conservationist. *Ecopsychology* 5, no. 2: 70–79.

Friedman, A. J. 2008. *Framework for evaluating impacts of informal science education projects: Report from a National Science Foundation workshop*. Arlington, VA: The National Science Foundation.

Gates, Christopher T. 2003. The civic landscape. *National Civic Review* 92, no. 1: 67–72.

George, Kelly A., Kristina M. Slagle, Robyn S. Wilson, Steven J. Moeller, and Jeremy T. Bruskotter. 2016. Changes in attitudes toward animals in the United States from 1978 to 2014. *Biological Conservation* 201: 237–242.

Gibson, Donald. 2014. *Ecology, ideology and power*. San Diego, CA: Progressive Press.

Goulart, Vinicius, Pedro Guimaraes de Azevedo, Joanna van de Schepop, Camila Teixeira, Luciana Barçante, Cristiano Schetini de Azevedo, and Robert John Young. 2009. Gaps in the study of zoo and wild animal welfare. *Zoo Biology* 28: 561–573.

Grajal, Alejandro, Jerry F. Luebke, Lisa-Anne DeGregoria Kelly, Jennifer Matiasek, Susan Clayton, Bryan T. Karazsia, Carol D. Saunders, Susan R. Goldman, Michael E. Mann, and Ricardo Stanoss. 2017. The complex relationship between personal sense of connection to animals and self-reported proenvironmental behaviors by zoo visitors. *Conservation Biology* 31, no. 2: 322–330.

Grammar, Karl, Berhnard Fink, and Nick Neave. 2005. Human pheromones and sexual attraction. *European Journal of Obstetrics & Gynecology and Reproductive Biology* 118, no. 2: 135–142.

Grandin, Temple. 1980. Observations of cattle behavior applied to the design of cattle-handling facilities. *Applied Animal Ethology* 6, no. 1: 19–31.

Grazian, David. 2015. *American zoos: A sociological safari*. Princeton, NJ: Princeton University Press.

Gupta, Rupanwita, John Voiklis, Shelley J. Rank, Joseph de la Torre Dwyer, John Fraser, Kate Flinner, and Kathryn Nock. 2020. Public perceptions of the STEM learning ecology – perspectives from a national sample in the US. *International Journal of Science Education*: https://doi.org/10.1080/21548455.2020.1719291.

Gwynne, J. A. 2007. Inspiration for conservation: Moving audiences to care. In Zimmermann et al. eds., *Zoos in the 21st century: Catalysts for conservation?* Cambridge: Cambridge University Press.

Halliday, M. A. K. 1993. Towards a language-based theory of learning. *Linguistics and Education* 5: 93–116.

Hamilton, Erin Miller, Meaghan L. Guckian, and Raymond De Young. 2018. Living well and living green: Participant conceptualizations of green citizenship. In Walter Leal Filho, Robert W. Marans, and John Callewaert, eds., *Handbook of sustainability and social science research*, 315–334. New York: Springer International Publishing.

Hancocks, David. 1995. An introduction to reintroduction. In Bryan G. Norton, Michael Hutchins, Elizabeth F. Stevens, and Terry L. Maple, eds., *Ethics on the ark: Zoos, animal welfare, and wildlife conservation*, 181–183. Washington, DC: Smithsonian Institution Press.

2003. *A different nature: The paradoxical world of zoos and their uncertain future*. Berkley: University of California Press.

Hanson, Elizabeth. 2002. *Animal attractions: Nature on display in American zoos*. Princeton, NJ: Princeton University Press.

Hayward, Jeff, and Marilyn Rothenberg. 2010. Measuring success in the Congo gorilla forest conservation exhibit. *Curator: The Museum Journal* 47, no. 3: 261–282.

Heath, Shirley Brice. 1983. *Ways with words: Language, life and work in communities and classrooms*. Cambridge: Cambridge University Press.

Heimlich, Joe E., and John H. Falk. 2009. Free-choice learning and the environment. In John H. Falk, Joe E. Heimlich, and Susan Foutz, eds., *Free-choice learning and the environment*, 11–21. Lantham, MD: AltaMira Press.

Hein, George E. 1998. *The constructivist museum*. New York: Routledge.

2004. John Dewey and museum education. *Curator: The Museum Journal* 47, no. 4: 413–427.

Heinrich, Carolyn, and Barbara A. Birney. 1992. Effects of live animal demonstrations on zoo visitors' retention of information. *Anthrozoos* 5, no. 2: 113–121.

Hemsworth, Paul H. 2003. Human–animal interactions in livestock production. *Applied Animal Behavior Science* 81, no. 3: 185–198.

Hemsworth, Paul H., and Grahame J. Coleman. 2010. *Human-livestock interactions: The stockperson and the productivity of intensively farmed animals*, second ed. Wallingford: CABI.

Herziger, Atar, Jana Berkessel, and Kamilla Knutsen Steinnes. 2020. Wean off green: On the (in)effectiveness of biospheric appeals for consumption curtailment. *Journal of Environmental Psychology*.

Herzog, Hal. 2010. *Some we love, some we hate, some we eat: Why it's so hard to think straight about animals*. New York: HarperCollins Publishers.

2011. The impact of pets on human health and psychological well-being: Fact, fiction, or hypothesis? *Current Directions in Psychological Science* 20, no. 4: 236–239.

Herzog, Harold A., and Lauren L. Golden. 2009. Moral emotions and social activism: The case of animal rights. *Journal of Social Issues* 65, no. 3: 485–498.

Herzog, Harold A., and Shelley Galvin. 1997. Common sense and the mental lives of animals: An empirical approach. In Robert W. Mitchell, Nicholas S. Thompson, and H. Lyn Miles, eds., *Anthropomorphism, anecdotes, and animals*, 237–253. Albany: State University of New York Press.

Holland, Maximilian. 2004. Social bonding & nurture kinship: Compatibility between cultural and biological approaches. Ph.D. diss., London School of Economics and Political Science.

Holtorf, Cornelius. 2018. The zoo as a realm of memory. *Anthropological Journal of European Cultures* 22, no. 1: 98–114.

Horvath, Kelsey, Dario Angeletti, Giuseppe Nascetti, and Claudio Carere. 2013. Invertebrate welfare: An overlooked issue. *Ann Ist Super Sanità* 49: 9–17.

Hosey, Geoff, and Vicky Melfi. 2012. Human–animal bonds between zoo professionals and the animals in their care. *Zoo Biology* 31, no. 1: 13–26.

2015. Are we ignoring neutral and negative human–animal relationships in zoos? *Zoo Biology* 34, no. 1: 1–8.

Huizinga, Johan. 1938. *Homo ludens: A study of the play-element in culture*. Brooklyn, NY: Angelico Press.

Human, Katy. 2005. The Bible as museum guide. *Denver Post*, September 24.

Hume, David. 1739/1978. *A treatise of human nature*. Oxford: Oxford University Press.

International Union for Conservation of Nature and Natural Resources. 1971. *IUCN yearbook: Annual report of the IUCN for the year*. Morges: IUCN.

Jamieson, Dale. 2002. *Morality's progress: Essays on humans, other animals, and the rest of nature*. New York: Clarendon Press.

Jensen, Eric. 2014. Evaluating children's conservation biology learning at the zoo. *Conservation Biology* 28, no. 4: 1004–1011.

Jensen, Eric, Andrew Moss, and Markus Gusset. 2017. Quantifying long-term impact of zoo and aquarium visits on biodiversity-related learning outcomes. *Zoo Biology* 36, no. 4: 294–297.

Kahn, Peter H., Jr. 1999. *The human relationship with nature.* Cambridge, MA: MIT Press.

2009. Please don't visit zoos! Zoos reify the human drive to dominate the Other. *Psychology Today.* www.psychologytoday.com/us/blog/human-nature/200910/please-dont-visit-zoos.

Kahn, Peter H., Jr., Carol Saunders, Rachel L. Severson, Olin Eugene Myers, Jr., and Brian T. Gill. 2008. Moral and fearful affiliations with the animal world: Children's conceptions of bats. *Anthrozoos* 21: 375–386.

Kahn, Peter H., Jr., and Stephen R. Kellert, eds. 2002. *Children and nature: Psychological, sociocultural, and evolutionary investigations.* Cambridge, MA: MIT Press.

Kaiser, Florian G., and P. Wesley Schultz. 2009. The attitude-behavior relationship: A test of three models of the moderating role of behavioral difficulty. *Journal of Applied Social Psychology* 39, no. 1: 186–207.

Kalof, Linda. 2000. The multi-layered discourses of animal concern. In Helen Addams and John L. R. Proops, eds., *Social discourse and environmental policy,* 174–195. Cheltenham: Edward Elgar Publishers.

Kals, Elisabeth. 1996. Are proenvironmental commitments moderated by health concerns or perceived justice? In Leo Mondata and Melvin J. Lerner, eds., *Current societal concerns about justice,* 231–258. New York: Plenum Press.

Karp, Ivan, Christine Mullen Kreamer, and Steven Levine, eds. 1992. *Museums and communities: The politics of public culture.* Washington, DC: Smithsonian Books.

Keitsch, Martina. 2018. Structuring ethical interpretations of the sustainable development goals: Concepts, implications and progress. *Sustainability* 10, no. 3: 829.

Kellert, Stephen R. 1980. American attitude toward and knowledge of animals: An update. *International Journal for the Study of Animal Problems* 1: 87–119.

1997. *The value of life: Biological diversity and human society.* Washington, DC: Island Press.

2012. *Birthright: People and nature in the modern world.* New Haven, CT: Yale University Press.

Kendall, Julie E., and Kenneth E. Kendall. 1993. Metaphors and methodologies: Living beyond the systems machine. *MIS Quarterly* 17, no. 2: 149–171.

Keynes, John Maynard. 1936. *The general theory of employment, interest and money.* London: Macmillan.

Kidd, Aline H., and Robert M. Kidd. 1996. Developmental factors leading to positive attitudes toward wildlife and conservation. *Applied Animal Behavior Science* 47, no. 1–2: 119–125.

Kinsley, David. 1995. *Ecology and religion: Ecological spirituality in cross-cultural perspective.* Englewood Cliffs, NJ: Prentice Hall.

Kisiel, James, Shawn Rowe, Melanie Ani Vartabedian, and Charles Kopczak. 2012. Evidence for family engagement in scientific reasoning at interactive animal exhibits. *Science Education* 96: 1047–1070.

Kisling, Vernon N., ed. 2000. *Zoo and aquarium history: Ancient animal collections to zoological gardens.* Boca Raton, FL: CRC Press.

Klenosky, David B. and Carol D. Saunders. 2008. Put me in the zoo! A laddering study of zoo visitor motives. *Tourism Review International* 11, no. 3: 317–327.

Korn, Randi. 2004. Self-portrait: First know thyself, then serve your public. *Museum News*, 33–52.

Kreger, Michael D., Michael Hutchins, and Nina Fascione. Context, ethics, and environmental enrichment in zoos and aquariums. In David J. Shepherdson, Jill D. Mellen, and Michael Hutchins, eds., *Second nature: Environmental enrichment for captive animals*, 59–82. Washington, DC: Smithsonian Institution Press.

Kyle, Gerard, Alan Graefe, and Robert Manning. 2005. Testing the dimensionality of place attachment in recreational settings. *Environment and Behavior* 37, no. 2: 153–177.

Lee, Keekok. 2005. *Zoos: A philosophical tour.* Basingstoke: Palgrave Macmillan.

Leinhardt, Gaea, and Karen Knutson. 2004. *Listening in on museum conversations.* Walnut Creek, CA: AltaMira Press.

Leopold, Aldo. 1970. *A Sand County almanac.* New York: Ballantine Books.

Levine, Peter. 2007. *The future of democracy: Developing the next generation of American citizens.* Medford, MA: Tufts University Press.

Lipsky, Laura van Dernoot. 2009. *Trauma stewardship: An everyday guide to caring for self while caring for others.* San Francisco, CA: Berrett-Heohler.

LoBue, Vanessa, Megan Bloom Pickard, Kathleen Sherman, Chrystal Axford, and Judy DeLoache. 2013. Young children's interest in live animals. *British Journal of Developmental Psychology* 31, no. 1: 57–69.

Longhurst, Brian, Gaynor Bagnall, and Mike Savage. 2004. Audiences, museums and the English middle class. *Museum and Society* 2, no. 2: 104–124.

Lovelock, James E. 1987. *Gaia: A new look at life on earth.* New York: Oxford University Press.

Lovelock, James E., and Lynn Margulis. 1974. Atmospheric homeostasis by and for the biosphere: The Gaia hypothesis. *Tellus* 26, no. 1–2: 2–10.

Luebke, Jerry F. Zoo exhibit experiences and visitors' affective reactions: A preliminary study. 2018. *Curator: The Museum Journal* 61, no. 2: 345–352.

Luebke, Jerry F., Jason V. Watters, Jan Packer, Lance J. Miller, and David M. Powell. 2016. Zoo visitors' affective responses to observing animal behaviors. *Visitor Studies* 19, no. 1: 60–76.

Luebke, Jerry F., and Jennifer Matiasek. 2013. An exploratory study of zoo visitors' exhibit experiences and reactions. *Zoo Biology* 32, no. 4: 407–416.

Lukas, Kristen E., and Stephen R. Ross. 2005. Zoo visitor knowledge and attitudes toward gorillas and chimpanzees. *Journal of Environmental Education* 36, no. 4: 33–48.

Lyng, Stephen. 2004. *Edgework: The sociology of risk-taking*. New York: Routledge.

MacDonald, Edith. 2015. Quantifying the impact of Wellington Zoo's persuasive communication campaign on post-visit behavior. *Zoo Biology* 34, no. 2: 163–169.

MacDonald, Edith, Taciano Milfont, and Michael Gavin. 2016. Applying the elaboration likelihood model to increase recall of conservation messages and elaboration by zoo visitors. *Journal of Sustainable Tourism* 24, no. 6: 866–881.

Mace, Georgina, Andrew Balmford, Nigel Leader-Williams, Andrea Manica, Olivia Walter, Chris West, and Alexandra Zimmerman. 2007. Measuring conservation success: Assessing zoos' contributions. In Alexandra Zimmermann, Matthew Hatchwell, Lesley Dickie, and Chris West, eds., *Zoos in the 21st century: Catalysts for conservation?* 322–342. Cambridge: Cambridge University Press.

Malamud, Randy. 1998. *Reading zoos: Representations of animals and captivity*. New York: New York University Press.

2003. *Poetic animals and animal souls*. London: Palgrave Macmillan.

Manfredo, Michael J. 2008. Integrating concepts: Demonstration of a multilevel model for exploring the rise of mutualism value orientations in post-industrial society. In Michael J. Manfedro, ed., *Who cares about wildlife? Social concepts for exploring human-wildlife relationships and conservation issues*, 191–217. New York: Springer.

Manfredo, Michael J., Tara L. Teel, and Alan D. Bright. 2003. Why are public values toward wildlife changing? *Human Dimensions of Wildlife* 8, no. 4: 287–306.

Manfredo, Michael J., Tara L. Teel, and Harry Zinn. 2009. Understanding global values toward wildlife. In Michael J. Manfredo, Jerry J. Vaske, Perry J. Brown, Daniel J. Decker, and Esther A. Duke, eds., *Wildlife and society: The science of human dimensions*, 31–43. Washington, DC: Island Press.

Manfredo, Michael J., Tara L. Teel, and Kimberly L. Henry. 2009. Linking society and environment: A multilevel model of shifting wildlife value orientations in the western U.S. *Social Science Quarterly* 90, no. 2: 407–427.

Mann-Lang, Judy, Roy Ballantyne, and Jan Packer. 2016. Does more education mean less fun? A comparison of two animal presentations. *International Zoo Yearbook* 50, no. 1: 155–164.

Manzo, Lynne C. 2003. Beyond house and haven: Toward a revisioning of emotional relationships with places. *Journal of Environmental Psychology* 23: 47–61.

Maor, Moshe. 2016. Emotion-driven negative policy bubbles. *Policy Sciences* 49, no. 2: 191–210.

Maple, Terry, Rita McManamon, and Elizabeth Stevens. 1995. Defining the good zoo: Animal care, maintenance, and welfare. In Bryan G. Norton, Michael Hutchins, Elizabeth F. Stevens, and Terry L. Maple, eds., *Ethics on the ark: Zoos, animal welfare, and wildlife conservation*, 219–234. Washington, DC: Smithsonian Institution Press.

Markowitz, Hal. 1981. *Behavioral enrichment in the zoo*. New York: Van Nostrand Rheinhold.

Martínez-Macipe, M., Céline Lafont-Lecuelle, Xavier Manteca, Patrick Pageat, and Alessandro Cozzi. 2015. Evaluation of an innovative approach for sensory enrichment in zoos: Semiochemical stimulation for captive lions (Panthera leo). *Animal Welfare* 24, no. 4: 455–461.

Mascia, Michael B., J. Peter Brosius, Tracy A. Dobson, Bruce C. Forbes, Leah Horowitz, Margaret A. McKean, and Nancy J. Turner. 2003. Editorial: Conservation and the social sciences. *Conservation Biology* 17: 649–650.

Mason, Peter. 2000. Zoo tourism: The need for more research. *Journal of Sustainable Tourism* 8, no. 4: 333–339.

Mather, Jennifer A. 2001. Animal suffering: An invertebrate perspective. *Journal of Applied Animal Welfare Science* 4, no. 2: 151–156.

Matiasek, Jennifer, and Jerry F. Luebke. 2014. Mission, messages, and measures: Engaging zoo educators in environmental education program evaluation. *Studies in Educational Evaluation* 41: 77–84.

McGrath, Joseph E., Holly Arrow, and Jennifer L. Berdahl. 2000. The study of groups: Past, present, and future. *Personality and Social Psychology Review* 4, no. 1: 95–105.

Mead, George Herbert. 1934. *Mind, self, and society*. Chicago: University of Chicago Press.

Melfi, Vicky. 2009. There are big gaps in our knowledge, and thus approach, to zoo animal welfare: A case for evidence-based zoo animal management. *Zoo Biology* 28: 574–588.

Melson, Gail F. 2001. *Why the wild things are: Animals in the lives of children*. Cambridge, MA: Harvard University Press.

Meyers, Ronald B. 2002. A heuristic for environmental values and ethics, and a psychometric instrument to measure adult environmental ethics and willingness to protect the environment. Ph.D. diss., The Ohio State University.

Morgan, Mark, and Marlana Hodgkinson. 1999. The motivation and social orientation of visitors attending a contemporary zoological park. *Environment and Behavior* 31, no. 2: 227–239.

Morgan, Kathleen N., Scott W. Line, and Hal Markowitz, 1998. Zoos, enrichment, and the skeptical observer: The practical value of assessment. In David J. Shepherdson, Jill D. Mellen, and Michael Hutchins, eds., *Second nature: Environmental enrichment for captive animals*, 153–171. Washington, DC: Smithsonian Institution Press.

Morris, Desmond. 1969/1994. *The human zoo: A zoologist's study of the urban animal*. New York: Vintage Books.

Moss, Andrew, Eric Jensen, and Markus Gusset. 2015. Evaluating the contribution of zoos and aquariums to Aichi Biodiversity Target 1. *Conservation Biology* 29, no. 2: 537–544.

Mullan, Bob, and Garry Marvin. 1987. *Zoo culture*. London: Weidenfeld and Nicolson.

Musschenga, Albert W. 2002. Naturalness: Beyond animal welfare. *Journal of Agricultural and Environmental Ethics* 15, no. 2: 171–186.

Myers, Jr., Olin Eugene. 2007. *The significance of children and animals: Social development and our connections to other species*, second ed. West Lafayette, IN: Purdue University Press.

Myers, Jr., Olin Eugene, and Carol D. Saunders. 2002. Animals as links toward developing caring relationships with the natural world. In Peter H. Kahn, Jr. and Stephen R. Kellert, eds., *Children and nature: Psychological, sociocultural, and evolutionary investigations*, 153–178. Cambridge, MA: MIT Press.

Myers, Jr., Olin Eugene, Carol D. Saunders, and Andrej A. Birjulin. 2004. Emotional dimensions of watching zoo animals: An experience sampling study building on insights from psychology. *Curator: The Museum Journal* 47, no. 3: 299–321.

Nash, Roderick. 1982. *Wilderness and the American mind*, third ed. New Haven, CT: Yale University Press.

Netzley, Patricia D. 1999. Environmental Literature: An Encyclopedia of Works, Authors, and Themes. Santa Barbara, CA: ABC-CLIO, LLC

Norgaard, Kari Marie. 2011. *Living in denial: Climate change, emotions, and everyday life*. Boston: MIT Press.

Norton, Bryan G., Michael Hutchins, Elizabeth F. Stevens, and Terry L. Maple, eds. 1995. *Ethics on the ark: Zoos, animal welfare and wildlife conservation*. Washington, DC: Smithsonian Institution Press.

Onyx, Jenny, and Paul Bullen. 2000. Measuring social capital in five communities. *The Journal of Applied Behavioral Science* 36, no. 1: 23–42.

Opotow, Susan. 2011a. Moral exclusion. In Daniel J. Christie, ed., *Encyclopedia of peace psychology*. Hoboken, NJ: John Wiley and Sons, Inc.

  2011b. Social injustice. In Daniel J. Christie, ed., *Encyclopedia of peace psychology*. Hoboken, NJ: John Wiley and Sons, Inc.

Oreg, Shaul, and Tally Katz-Gerro. 2006. Predicting proenvironmental behavior cross-nationally: Values, the theory of planned behavior, and value-belief-norm theory. *Environment and Behavior* 38, no. 4: 462–483.

Osborn, Jr., Fairfield, ed. 1944. *The Pacific world*. New York: W. W. Norton & Company.

  1948. *Our plundered planet*. London: Faber and Faber.

  1953. *The limits of the earth*. London: Faber and Faber.

  ed. 1962. *Our crowded planet: Essays on the pressures of population*. Garden City, NY: Doubleday & Company, Inc.

Packer, Jan. 2006. Learning for fun: The unique contribution of educational leisure experiences. *Curator: The Museum Journal* 49, no. 3: 329–344.

Packer, Jan, and Roy Ballantyne. 2004. Is educational leisure a contradiction in terms? Exploring the synergy of education and entertainment. *Annals of Leisure Research* 7, no. 1: 54–71.

2010. The role of zoos and aquariums in education for a sustainable future. *New Directions for Adult and Continuing Education*, no. 127: 25–34.

2012. Comparing captive and non-captive wildlife tourism. *Annals of Tourism Research* 39, no. 2: 1242–1245.

Packer, Martin, Norma Haan, Paola Theodorou, and Gary Yabrove. 1985. Moral action of four-year-olds. In Norma Haan, Eliane Aerts, and Bruce A. B. Cooper, eds., *On moral grounds: The search for practical morality*, 276–305. New York: New York University Press.

Panksepp, Jaak. 2011. Cross-species affective neuroscience decoding of the primal affective experiences of humans and related animals. *PLoS ONE* 6, no. 8: e21236.

Pearson, Elissa L., Rachel Lowry, Jillian Dorrian, and Carla A. Litchfield. 2014. Evaluating the conservation impact of an innovative zoo-based educational campaign: "Don't palm us off" for orang-utan conservation. *Zoo Biology* 33, no. 3: 184–196.

Pekarik, Andrew J. 2004. Eye-to-eye with animals and ourselves. *Curator: The Museum Journal* 47, no. 3: 257–260.

Pekarik, Andrew J., Zahava D. Doering, and David A. Karns. 1999. Exploring satisfying experiences in museums. *Curator: The Museum Journal* 42, no. 2: 117–129.

Persson, Tomas, Gabriela-Alina Sauciuc, and Elainie Alenkær Madsen. 2018. Spontaneous cross-species imitation in interactions between chimpanzees and zoo visitors. *Primates* 59, no. 1: 19–29.

Podilchak, Walter. 1991. Distinctions of fun, enjoyment and leisure. *Leisure Studies* 10, no. 2: 133–148.

Polakowski, Kenneth J. 1987. *Zoo design: The reality of wild illusions*. Ann Arbor: The University of Michigan School of Natural Resources.

1989. A design approach to zoological exhibits: The zoo as theater. *Zoo Biology* 8, no. S1: 127–139.

Powell, Robert, and Sam H. Ham. 2008. Can ecotourism interpretation really lead to pro-conservation knowledge, attitudes and behavior? Evidence from the Galapagos islands. *Journal of Sustainable Tourism* 16: 467–489.

Preston, Jane, ed. 2013. *The language of conservation: Poetry in library and zoo collaborations*. New York: Poets House.

Proshansky, Harold M. 1978. The city and self-identity. *Environment and Behavior* 10, no. 2: 147–169.

Proshansky, Harold M., Abbe K. Fabian, and Robert Kaminoff. 1983. Place identity. *Journal of Environmental Psychology* 3: 57–83.

Rabb, George B. 2004. The evolution of zoos from menageries to centers of conservation and caring. *Curator: The Museum Journal* 47, no. 3: 237–246.

Rabb, George B., and Carol D. Saunders. 2006. The future of zoos and aquariums: Conservation and caring. *International Zoo Yearbook* 39, no. 1: 1–26.

Rank, Shelley J., John Voiklis, Rupanwita Gupta, John Fraser, and Kate Flinner. 2018. Understanding organizational trust of zoos and aquariums. In Kathleen P. Hunt, ed., *Understanding the role of trust and credibility in science communication*: https://doi.org/10.31274/sciencecommunication-181114-16.

Ranson, Stewart. 2003. Public accountability in the age of neo-liberal governance. *Journal of Education Policy* 18, no. 5: 459–480.

Rasmussen, L. E. L., Terry D. Lee, Aijun Zhang, Wendell L. Roelofs, and G. Doyle Daves. 1997. Purification, identification, concentration and bio-activity of (Z)-7-dodecen-1-yl acetate: Sex pheromone of the female Asian elephant, Elephas maximus. *Chemical Senses* 22, no. 4: 417–437.

Reade, Louise S., and Natalie K. Waran. 1996. The modern zoo: How do people perceive zoo animals? *Applied Animal Behaviour Science* 47: 109–118.

Regan, Tom. 1983. *The case for animal rights*. Berkeley: University of California Press.

Robison, Lindon J., A. Allan Schmid, and Marcelo E. Siles. 2002. Is social capital really capital? *Review of Social Economy* 60, no. 1: 1–21.

Roggenbuck, Joseph W., and B. L. Driver. 2000. Benefits of nonfacilitated uses of wilderness. In Stephen F. McCool, David N. Cole, William T. Borrie, and Jennifer O'Loughlin, eds., *Wilderness science in a time of change conference volume 3: Wilderness as a place for scientific inquiry*, 33–49. Ogden, UT: U.S. Department of Agriculture, Forest Service, Rocky Mountain Research Station.

Röska-Hardy, Louise S., and Eva M. Neumann-Held, eds. 2009. *Learning from animals? Examining the nature of human uniqueness*. New York: Psychology Press.

Røskaft, Eivin, Tore Bjerke, Bjørn Kaltenborn, John D. C. Linnell, and Reidar Andersen. 2003. Patterns of self-reported fear towards large carnivores among the Norwegian public. *Evolution and Human Behavior* 24, no. 3: 184–198.

Roszak, Theodore, Mary E. Gomes, and Allen D. Kanner, eds. 1995. *Ecopsychology: Restoring the earth, healing the mind*. San Francisco: Sierra Club Books.

Rothfels, Nigel. 2002. *Savages and beasts: The birth of the modern zoo*. Baltimore, MD: Johns Hopkins University Press.

Rounds, Jay. 2006. Doing identity work in museums. *Curator: The Museum Journal* 49, no. 2: 133–150.

Rowlands, Mark. 2002. *Animals like us*. London: Verso.

Rudolph, John L., and Shusaku Horibe. 2015. What do we mean by science education for civic engagement? *Journal of Research in Science Teaching* 53, no. 6: 805–820.

Ryan, Chris, and Jan Saward. 2004. The zoo as ecotourism attraction-visitor reactions, perceptions and management implications: The case of Hamilton Zoo, New Zealand. *Journal of Sustainable Tourism* 12, no. 3: 245–266.

Saunders, Carol D. 2003. The emerging field of conservation psychology. *Human Ecology Review* 10, no. 2: 137–149.

2008. Double-edged swords? Collective identity and solidarity in the environment movement. *British Journal of Sociology* 59, no. 2: 227–253.

Saunders, Carol D., John Fraser, and Ronald Meyers. 2005, October. Applying a psychometric instrument to explore the range of environmental ethical beliefs held by zoo visitors. Paper presented at the Annual Society for Human Ecology Conference, Salt Lake City, UT.

Saunders, Carol D., and Olin Eugene Myers, Jr. 2003. Exploring the potential of conservation psychology. *Human Ecology Review* 10, no. 2: 1–3.

Schultz, P. Wesley. 2000. Empathizing with nature: The effects of perspective taking on concern for environmental issues. *Journal of Social Issues* 56, no. 3: 391–406.

2002. Inclusion with nature: Understanding the psychology of human-nature interactions. In Peter Schmuck and P. Wesley Schultz, eds., *The psychology of sustainable development*, 61–78. New York: Kluwer.

Schultz, P. Wesley, Jessica M. Nolan, Robert B. Cialdini, Noah J. Goldstein, and Vladas Griskevicius. 2007. The constructive, destructive, and reconstructive power of social norms. *Psychological Science* 18: 429–434.

Schultz, P. Wesley, and Jennifer Tabanico. 2007. Self, identity, and the natural environment. *Journal of Applied Social Psychology* 37, no. 6: 1219–1247.

Schultz, P. Wesley, and Lynnette Zelezny. 1999. Values as predictors of environmental attitudes, evidence for consistency across 14 countries. *Journal of Environmental Psychology* 19, no. 1: 255–265.

Schwartz-DuPre, Rae Lynn, and Helen Morgan Parmett. 2017. Curious about George: Postcolonial science and technology studies, STEM education policy, and colonial iconicity. *Textual Practice* 32, no. 4: 707–725.

Seligman, Martin E. P. 2002. *Authentic happiness: Using the new positive psychology to realize your potential for lasting fulfillment.* New York: Free Press.

Shackley, Myra. 1996. *Wildlife tourism.* London: Routledge.

Shepherdson, David J., Jill D. Mellen, and Michael Hutchins, eds. 1998. *Second nature: Environmental enrichment for captive animals.* Washington, DC: Smithsonian Institution.

Shepherdson, David, Karen D. Lewis, Kathy Carlstead, Joan Bauman, and Nancy Perrin. 2013. Individual and environmental factors associated with stereotypic behavior and fecal glucocorticoid metabolite levels in zoo housed polar bears. *Applied Animal Behaviour Science* 147: 268–277.

Shiota, Michelle N., Dacher Keltner, and Amanda Mossman. 2007. The nature of awe: Elicitors, appraisals, and effects on self-concept. *Cognition and Emotion* 21, no. 5: 944–963.

Sickler, Jessica, and John Fraser. 2009. Enjoyment in zoos. *Leisure Studies* 28, no. 3: 313–331.

Sickler, Jessica, John Fraser, Sarah Gruber, Paul Boyle, Tom Webler, and Diana Reiss. 2006. *Thinking about dolphins thinking (WCS working paper no. 27).* New York: Wildlife Conservation Society.

Singer, Peter. 1975. *Animal liberation: The definitive classic of the animal movement*. New York: HarperCollins Publishers.

Skibins, Jeffrey C., and Robert B. Powell. 2013. Conservation caring: Measuring the influence of zoo visitors' connection to wildlife on pro-conservation behaviors. *Zoo Biology* 32, no. 5: 528–540.

Sobel, David. 1996. *Beyond ecophobia: Reclaiming the heart in nature education*. Great Barrington, MA: The Orion Society and The Myrin Institute.

Soulé, Michael E. 1985. What is conservation biology? *Bioscience* 25: 727–734.

———. 1986. Conservation biology and the "real world." In Michael E. Soulé, ed., *Conservation biology: The science of scarcity and diversity*, 1–12. Sunderland, MA: Sinauer Associates.

———. 1987. History of the society for conservation biology: How and why we got here. *Conservation Biology* 1, no. 1: 4–5.

Spitzer, William, and John Fraser. 2020. Advancing community science literacy. *Journal of Museum Education* 45, no. 1: 5–15.

Spitzer, William, John Fraser, Julie Sweetland, and John Voiklis. 2020. Applied social science to scale climate communications impact. In Joseph Henderson and Andrea Drewes, eds., *Teaching climate change in the United States*, 123–142. New York: Routledge.

Stein, Gertrude. 1993. Identity a poem. In Ulla E. Dydo, ed., *A Stein reader*. Evanston, IL: Northwestern University Press.

Stern, Daniel N. 1985. *The interpersonal world of the infant*. New York: Basic Books.

Stern, Paul C. 2000. Toward a coherent theory of environmentally significant behavior. *Journal of Social Issues* 56, no. 3: 407–424.

Stern, Paul C., and Thomas Dietz. 1994. The value basis of environmental concern. *Journal of Social Issues* 50: 65–84.

Stern, Paul C., Thomas Dietz, Troy Abel, Gregory A. Guagnano, and Linda Kalof. 1999. A value-belief-norm theory of support for social movements: The case of environmentalism. *Human Ecology Review* 6, no. 2: 81–97.

Strauss, Anselm, and Juliet Corbin. 1998. *Basics of qualitative research: Techniques and procedures for developing grounded theory*, second ed. Thousand Oaks, CA: Sage Publications.

Stryker, Sheldon, and Richard T. Serpe. 1994. Identity salience and psychological centrality: Equivalent, overlapping, or complementary concepts? *Social Psychology Quarterly* 57, no. 1: 16–35.

Sutherland, Anne, and Jeffrey E. Nash. 1994. Animal rights as a new environmental cosmology. *Qualitative Sociology* 17, no. 2: 171–186.

Swanagan, Jeffrey S. 2000. Factors influencing zoo visitors' conservation attitudes and behavior. *Journal of Environmental Education* 31, no. 4: 26–31.

Swim, Janet K., and Brittany Bloodhart. 2013. Admonishment and praise: Interpersonal mechanisms for promoting pro-environmental behavior. *Ecopsychology: Special Section on Confronting Unsustainable Behaviors* 5: 23–35.

Swim, Janet K., and John Fraser. 2013. Fostering hope in climate change educators. *Journal of Museum Education* 38, no. 3: 286–297.

———. 2014. Zoo and aquarium professionals' concerns and confidence about climate change education. *Journal of Geoscience Education* 62, no. 3: 495–501.

Swim, Janet K., John Fraser, and Nathaniel Geiger. 2014. Teaching the choir to sing: Use of social science information to promote public discourse on climate change. *Journal of Land Use and Environmental Law* 30, no. 1: 91–117.

Swim, Janet K., Nathaniel Geiger, John Fraser, and Nette Pletcher. 2017. Climate change education at nature-based museums. *Curator: The Museum Journal* 60, no. 1: 101–119.

Tajfel, Henri. 1982. Social psychology of intergroups relations. *Annual Review of Psychology* 33: 1–39.

Tajfel, Henri, and John C. Turner. 1986. The social identity theory of intergroup behavior. In John T. Jost and Jim Sidanius, eds., *Key readings in social psychology*, 276–293. New York: Psychology Press.

Tedlock, Dennis, and Bruce Mannheim, eds. 1995. *The dialogic emergence of culture*. Urbana: University of Illinois Press.

Thomashow, Mitchell. 1995. *Ecological identity: Becoming a reflective environmentalist*. Cambridge, MA: MIT Press.

———. 2002. *Bringing the biosphere home, learning to perceive global environmental change*. Cambridge, MA: MIT Press.

Tirindelli, Roberto, Michele Dibattista, Simone Pifferi, and Anna Menini. 2009. From pheromones to behavior. *Physiology Review* 89, no. 3: 921–956.

Tofield, Sara, Richard K. Coll, Brent Vyle, and Rachel Bolstad. 2003. Zoos as a source of free choice learning. *Research in Science & Technological Education* 21, no. 1: 67–100.

Tomas, Stacy R., John Crompton, and David Scott. 2003. Assessing service quality and benefits sought among zoological park visitors. *Journal of Park and Recreation Administration* 21, no. 52: 105–124.

Tunnicliffe, Susan Dale. 1995. Talking about animals: Studies of young children visiting zoos, a museum and a farm. Ph.D. diss., King's College, University of London.

Turley, Sophie K. 2001. Children and the demand for recreational experiences: The case of zoos. *Leisure Studies* 20, no. 1: 1–18.

Vonnegut, Kurt. 2014. *Welcome to the monkey house: The special edition*. New York: Random House.

Vygotsky, Lev S. 1978. *Mind in society: The development of higher psychological processes*. Cambridge, MA: Harvard University Press.

Waal, Frans B. M. 2019. *Mama's last hug: Animal emotions and what they tell us about ourselves*. New York: W. W. Norton & Company.

Wagner, Frederic H, and Henry C. Kenski. 1988. *Predator control and the sheep industry: The role of science in policy formation*. Claremont, CA: Regina Books.

Wagner, Kathy, John Falk, Cynthia Vernon, Jackie Ogden, Emily Routman, and Keith Winsten. 2006. *Why zoos and aquariums matter: Results from the multi-institutional research program*. Silver Spring, MD: Association of Zoos and Aquariums.

Weber, Max. 1905/2003. *The protestant ethic and the spirit of capitalism*. Trans. Talcott Parsons. Mineola, NY: Dover Publications.

Wenger, Étienne. 1998. *Communities of practice: Learning, meaning, and identity*. Cambridge: Cambridge University Press.

Wenger, Étienne, Beverly Trayner, and Maarten De Laat. 2011. *Promoting and assessing value creation in communities and networks: A conceptual framework*. Heerlen: Ruud de Moor Centrum.

Williams, Florence. 2017. *Nature fix: Why nature makes us happier, healthier, and more creative*. New York: W. W. Norton & Company.

Wilson, Alexander. 1992. *The culture of nature: North American landscape from Disney to the Exxon Valdez*. Cambridge, MA: Blackwell.

Wilson, E. O. 1984. *Biophilia: The human bond with other species*. Cambridge, MA: Harvard University Press.

1993. Biophilia and the conservation ethic. In Stephen R. Kellert and E. O. Wilson, eds., *The biophilia hypothesis*, 31–41. Washington, DC: Island Press.

Wood, Nichola, and Louise Waite. 2011. Editorial: Scales of belonging. *Emotion, Space and Society* 4, no. 4: 210–202.

Woods, Barbara. 2002. Good zoo/bad zoo: Visitor experiences in captive settings. *Anthrozoös* 15, no. 4: 343–360.

World Association of Zoos and Aquariums (WAZA). 2005. *The world zoo and aquarium conservation strategy: Building a future for wildlife*. Bern: World Association of Zoos and Aquariums.

2006. *Zoos and aquariums of the world*. Bern: World Association of Zoos and Aquariums.

2015. *Committing to conservation: The world zoo and aquarium conservation strategy*. Gland: World Association of Zoos and Aquariums.

Yerke, R. and A. Burns. 1991. Measuring the impact of animal shows on visitor attitudes. American Association of Zoological Parks and Aquariums 1991 Annual Conference Proceedings, 532–539. San Diego, CA: American Association of Zoological Parks and Aquariums.

Yero, Judith Lloyd. 2002. *Metaphors in education*. Retrieved from https://meandjulio.blog/wp-content/uploads/2018/07/Metaphors.pdf.

Zavetoski, Stephen. 2003. Constructing and maintaining ecological identities: The strategies of deep ecologists. In Susan Clayton and Susan Opotow, eds., *Identity and the natural environment: The psychological significance of nature*, 297–315. Cambridge, MA: MIT Press.

Zimmerman, Alexandra, Matthew Hatchwell, Lesley Dickie, and Chris West, eds. 2007. *Zoos in the 21st century: Catalysts for conservation?* Cambridge: Cambridge University Press.

Index

anthropomorphism, 47, 55, 59, 91, 111
Acampora, Ralph, 56, 59, 70
accountability, 15, 24, 85
affiliation, 16, 64, 99, 106, 108, 123,
  164
ambassador animals, 81–82
American Association of Museums, xi
American Association of Zoological Parks and
  Aquariums. *See* Association of Zoos and
  Aquariums
animal care staff. *See* keepers
animal caregivers. *See* keepers
anthropomorphism, 47, 55, 59, 91, 111
anthropomorphism, 47, 55, 59, 91, 111
anti-zoo critique, 24, 37, 111, 167, 181
Arizona-Sonora Desert Museum, 22
Association of Zoos and Aquariums (AZA), xi, 1,
  5, 24–26, 54, 58, 81, 125, 173
aversion, 83, 111
awe, 16, 34, 67–68, 102, 120–122, 129,
  180

Bandura, Albert, 153
belonging, 16, 108, 115, 118–123, 141–142,
  162, 171, 181
biophilia, 16, 108–118, 133, 143
bonding, 72, 165
bonding, human social, 15, 118–119
bridging, 97
British and Irish Association of Zoos and
  Aquariums, 26
Bronx Zoo, 7–10, 44, 46, 66, 80–81, 87, 90–91,
  116–118, 140, 157
Brookfield Zoo, 19, 47, 77, 131

Calgary Zoo, 101
Carr, David, xi, 29–30, 89, 154
community-as-classroom, 101
conservation psychology, 3, 11–13, 20–21, 23,
  59, 131, 150, 161
Central Park Zoo, 69, 94

character virtues, 66–68, 76
Chicago Zoological Society, 127, 131
collective action, 2–3, 28, 41, 90, 160, 168–176,
  181, 184
collective identity, 17, 123, 142, 158, 161–166,
  171, 173, 176
Colorado Springs Zoo, 144
Columbus Zoo and Aquarium, 45
connectedness, 7, 10, 16–17, 34, 92, 100, 103,
  113, 116–120, 142, 161, 163, 172
conservation biology, 13, 19–22, 84, 167
conservation education, 5, 14, 17, 22–23, 27,
  30, 37, 41, 47, 54, 63, 70, 103, 105, 160,
  167
conservation ethic, 16–17, 23, 33, 36, 41, 91,
  98, 101, 104, 116, 120, 130, 147, 155,
  178, 181
conservation movement, 5, 13, 148, 155,
  184
conservation psychology, 3, 11–13, 20–21, 23,
  59, 131, 150, 161
conservation values, 3, 8–9, 16, 20, 22, 106,
  141, 146–147, 184
continuity, 16, 99, 108–109, 112, 115, 117,
  119, 125, 138, 146, 183
COVID-19, xiv, 183
Csikszentmihalyi, Mihaly, 68–69, 119

democracy, 4, 89, 97
Denver Zoo, 144
dialogue, xiii, 9, 11, 15, 41, 44, 50, 56, 84–91,
  93–94, 96–97, 102, 137, 154, 160, 170,
  179, 184–185
Dierking, Lynn, 30–31
discourse, 6, 12, 14–15, 38, 40, 46–47, 65, 82,
  85–92, 115, 125, 148, 151, 163, 165, 175,
  179
dominion, xii, 48, 110, 113, 115–116

Egan, Kieran, 101–102
elaboration likelihood model, 93

emotion
  emotional experience, 15, 30, 40, 85, 99, 104,
    125–127, 131, 159
  emotional response, 14, 16, 74, 83, 102, 112,
    114, 116, 120, 124, 130
empathy, 16, 19, 21, 32, 40, 68, 77, 83, 100,
  112, 116, 121, 124, 128, 130, 153
environmental education, 8, 11, 18, 30, 104,
  106, 142, *See also* conservation education
environmental ethic, 6, 65, 147, 160, *See also*
  conservation ethic
environmental identity, 16, 103, 139, 141–142,
  144, 146, 150, 162, 164–165
environmental immersion, 39–40
environmental movement, 46–47, 70, 109,
  147–148, 175, *See also* conservation
  movement
environmental values, 60, 100, 105, 107, 141,
  163–164, *See also* conservation values
European Association of Zoos and Aquariums
  (EAZA), 1, 125

Falk, John H., 11, 30–31, 135
fear, 53–54, 68, 75, 78–81, 93, 105, 111, 116,
  124, 136–137, 156, 185
Fine, Gary, 64, 170
Frameworks Institute, 159
Fraser, John, xiii, 24–26, 32, 71, 80, 100–101,
  167, 170–171
free-choice learning, 11, 30, 37, 152
fun, 14, 57, 63–65, 73–74, 83, 91, 103–105,
  157, 163, 172

Gaia hypothesis, 109
group identity, 103, 161, *See also* collective
  identity
Gupta, Rupanwita, 168

Hagenbeck, Carl, 7, 35
Harrington, Brooke, 170
Heimlich, Joe E., 11, 31, 34

Jardin des Plantes, 4

Kahn, Peter H., Jr., 34, 59, 112–113, 115,
  119
Kalof, Linda, 115, 165
keepers, 24, 42, 48, 53–56, 93, 104, 125–127
Kellert, Stephen, 57, 110–111, 113–114, 120
Knology, xiii, 49, 159

learning outcomes, 5, 25, 32, 61, 68, 70, 85, 185
Lee, Keekok, 57–58, 70
legitimacy, zoo form, xii, 5, 9–10, 14, 26–27, 42,
  47–48, 70, 92, 98, 145

leisure, xiv, 14, 151, 156
  educational leisure, 14, 71–77
London Zoo, 14, 45, 61
London Zoological Society, 180

meaning making, 3, 14–15, 26, 30, 33, 40, 46,
  57, 64, 68–69, 82, 87, 91, 94, 103, 115,
  168, 178, 181
Melbourne Zoo, 157
metaphor, 14–15, 27, 39, 47–48, 53–54, 60,
  91–95, 100, 119, 137
Milwaukee Zoo, 22
morality, 10, 13, 17, 33–34, 41, 45, 53, 55, 93,
  121, 155
Myers, Olin Eugene, Jr., 27, 34, 120

National Network for Ocean and Climate
  Change Interpretation (NNOCCI), xiii,
  159, 174–178
New England Aquarium, 127, 159
New York Zoological Society, 6–10, 22, 46–47,
  86, 90, 180, *See also* Wildlife Conservation
  Society
New York Zoos and Aquarium, xiii, 143, 167

Oregon Zoo, xiii, 103, 122
Osborn, Henry Fairfield, Jr., 10, 12, 22, 50, 86,
  90, 142, 176

Panksepp, Jaak, 77–78, 125
peace psychology, 153
pedagogy, 11, 14, 36, 78, 110
Peterson, Christopher, 66
place attachment, 139–141
pleasure, 103, 106, 181, 185
Proshansky, Harold, 139

recreation, 4–5, 14, 24, 32, 40, 57, 63–65,
  70–72, 130
Regent's Park, 34
religion, identity and framework connections,
  21, 44, 65, 88, 107, 156
responsibility, 9–10, 20, 22, 24, 33, 59, 86, 90,
  99, 138, 144, 151, 180

San Diego Zoo, 22, 113, 118, 140
Saunders, Carol, 19–20, 131–132
self-concept, 112, 117, 120, 122, 137–139,
  141–142, 145, 161–162
Seligman, Martin, 66
Sickler, Jessica, 25–26, 71
Singapore Zoo, 140
Smithsonian National Zoo, 46, 140,
  180
social capital, 97–99, 161

social change, 11–12, 17, 24, 148, 151, 162, 170, 177
social norms, 12, 20, 28, 137, 145, 150, 152, 157–158, 170, 172
Society for Conservation Biology, 20, 22
Stein, Gertrude, 134
STEM (science, technology, engineering, and math) learning, 22, 32–33, 82, 92, 160, 167–168
Sustainable Development Goals, 10
Swanagan, Jeff, 45
Switzer, Tawnya, xiii

theory of planned behavior (TPB), 148–151
tiny publics, 160, 170, 175, 178
Toronto Zoo, xiii
trust, 15, 17, 24, 33, 42, 49, 97–98, 107, 121, 125, 177, 183

US Department of Agriculture, xi
United Nations, 10, 18

value orientations, 110–115, 154
value-action gap, 146

value-belief-norm (VBN) theory, 148–150, 171
Vancouver Aquarium, 101
Voiklis, John, 168
volunteers, xi, 12, 32–33, 61, 90, 98, 103, 105–106, 116, 155, 170–173, 181–182, 184
Vonnegut, Kurt, 92, 95

Why Zoos and Aquariums Matter (WZAM), 24–26, 31–34, 41–42, 60, 69, 73–75, 77–78, 89, 93, 98–101, 105–106, 142–144, 152–155
Wildlife Conservation Society, xiii, 7, 25, 116, 121, 173, *See* also New York Zoological Society
Wilson, E. O., 108–110, 112, 114–116
Woods Hole Oceanographic Institution, 159
World Zoo and Aquarium Association (WAZA), 1, 5, 12, 125
worldview analysis, 115–116

zoo design, 7, 23, 35–36, 38, 48, 79, 180
Zoological Gardens of London, 4, *See* also Regent's Park
zoonotic disease, xiv, 183–184

For EU product safety concerns, contact us at Calle de José Abascal, 56–1°,
28003 Madrid, Spain or eugpsr@cambridge.org.

www.ingramcontent.com/pod-product-compliance
Ingram Content Group UK Ltd.
Pitfield, Milton Keynes, MK11 3LW, UK
UKHW020453240426
470322UK00016B/328